The Anatomy of the Human Skel
A Question and Answer Tutorial Te:.

To Jean, David and Rosemary

The Anatomy of the Human Skeleton
A Question and Answer Tutorial Text

Eric E. Denman MA FRCS (Edin and Eng)

Examiner in Anatomy, FRCS (Edin); Formerly Consultant Accident and Orthopaedic Surgeon, Princess Margaret Hospital, Swindon, UK

CHURCHILL LIVINGSTONE
EDINBURGH LONDON MADRID MELBOURNE NEW YORK AND TOKYO 1992

CHURCHILL LIVINGSTONE
Medical Division of Longman Group UK Limited

Distributed in the United States of America by Churchill
Livingstone Inc., 650 Avenue of the Americas, New York,
N.Y. 10011 and by associated companies, branches and
representatives throughout the world.

First published 1992

ISBN 0 443-04498-8

British Library Cataloguing in Publication Data
A catalogue record for this book is available from the British
Library.

Library of Congress Cataloging in Publication Data
A catalog record for this book is available from the Library of
Congress

For Churchill Livingstone
Publisher: Simon Fathers
Editorial Co-ordination: Editorial Resources Unit
    Copy Editor: Jennifer Guise
Production Controller: Neil A. Dickson
Design: Design Resources Unit
Sales Promotion Executive: Louise Johnstone

Produced by Longman Singapore Publishers (Pte) Ltd
Printed in Singapore

Preface

In addition to the considerable and ever-increasing pressure of work and responsibility that young hospital doctors are exposed to, those who seek a specialist career must take on, in their own time and largely at their own expense, the extra task of studying for higher qualifications. The drive, energy, dedication and sheer hard work required to pass these examinations is seldom appreciated outside, and sometimes inside, the hospital environment. Often only the spouses are truly aware of the enormity of the task! Any help that can be offered to ease the path to success of these trainees can only be welcomed. Whether this small book fulfils its intended role in this respect can only be judged by them.

The following pages are directed at doctors whose interest guides them to surgery involving the skeleton. Mostly these doctors will be orthopaedic surgeons, but there is little modern surgery that does not involve some knowledge of bones and their anatomy. ENT and faciomaxillary surgeons, and neurosurgeons, must all have an intimate understanding of their fields in the skull; general and thoracic surgeons must know the chest wall and obstetricians the bony pelvis. Even plastic surgeons now regularly breach the periosteum!

With the reorganisation of the examinations of the UK and Irish Colleges of Surgeons, a specialist 'exit' examination has been introduced to follow, at the appropriate time, what was hitherto the basic science and general surgical parts of an 'entry', but final, qualification. In these new, later examinations the fields of knowledge are relatively narrower, but at the same time deeper. This must hold true of relevant anatomy, including the study of the skeleton. While most of the material in the following pages will interest the student of the basic science part of the first part (Part A) of the new examinations, some of it is beyond what can reasonably be expected of candidates at this stage. Candidates for the specialist examinations should, however, have mastered the field of their craft to the greater detail described.

It is obvious, therefore, that this text is not a primary work of instruction. Indeed it will be unintelligible as such. Readers are assumed to have a basic knowledge of anatomy up to at least a rusting graduate level. This book can then be used for revision purposes, after re-reading the standard student texts, of which there is an abundance of excellence. It is also assumed that the reader has access to anatomical atlases, the dry bones, and the X-ray films which bulge the notes in clinics and wards. It has not, unfortunately, been possible to include the many diagrams, reproductions of X-ray films, and CT scans that would have been helpful in description, and enlivening for the text, since this would have made the cost of production excessive.

The notes which comprise the text are the result of many happy hours of tutorials with the aspiring young surgeons of the Accident and Orthopaedic Department of the Princess Margaret Hospital, Swindon, and with the junior demonstrators of the Anatomy School of Oxford University, with whom it has been a privilege to be associated.

The value of the tutorial system lies in the reinforcement of knowledge it can provide by the persistent questioning it encourages, of the tutor of the students, of the students of the tutor, and of the tutor and the students of themselves. Often only by such exposure can the subject be made clearer, and retained in the memory for longer.

The tutorial format, of question and answer, presented in the text has both advantages and disadvantages. Among the advantages is the ability to highlight specific aspects of the subject, the importance of which is often not appreciated on first reading of fuller texts. Questions raised tend not to go away and nag at the mind. Thinking through the answers consolidates understanding and with repetition refreshes the memory. They probe areas of weakness that may not be discovered until too late across the viva table. They may pre-empt the examiner! The main disadvantage, of course, is the inevitably disjointed presentation they make when put together in an apparently continuous text. However, given the level of knowledge the reader already possesses, this should not be a problem.

The book is formed of two sections. In the first some basic histological topics are presented, since they are essential to the understanding of the pathology of the skeleton. In the second section the individual bones are covered in sequence.

Two idiosyncrasies of terminology must be explained. The term anulus is used in place of annulus, in support of the assumption that the word is derived from the Latin word anulus, meaning a ring, rather than from the word annus meaning a circuit, especially of the sun. The scientific term radiograph has been shunned for the more homely word X-ray film. Never in my career have I, or to my knowledge even my radiological colleagues, called for a nurse to produce the radiographs. She would probably think me rather pompous if I did!

It was not thought appropriate to include a full range of references, but a list of source books and suggested reading is appended at the end of the text.

I would like to record my thanks to Professor R.W. Guillery, head of the Department of Anatomy of Oxford University, for his great kindness to me, his assistance of my efforts, and, indeed, his courage in letting loose a district surgeon on the junior staff of his department. I wish also to thank my former consultant surgical colleagues at the Princess Margaret Hospital, Swindon, Mr Vivian Jones, for advice on the obstetric aspects of the pelvis, and Mr John Fieldhouse, for advice on the orbit, face and mandible. I am particularly indebted to Dr John Morris of the Anatomy Department, Oxford, for reading the script and making many valuable comments. Lastly, but most importantly, this book would not have been started, let alone completed, without the constant help and encouragement of my wife, Jean.

Marlborough, 1992   E.E.D.

THE ELEPHANT'S CHILD

I keep six honest serving-men
(They taught me all I knew);
Their names are What and Why and When
And How and Where and Who.
I send them over land and sea,
I send them east and west;
But after they have worked for me,
I give them all a rest.

I let them rest from nine till five,
For I am busy then,
As well as breakfast, lunch, and tea,
For they are hungry men.
But different folk have different views.
I know a person small —
She keeps ten million serving-men,
Who get no rest at all!

She sends 'em abroad on her own affairs,
From the second she opens her eyes —
One million Hows, two million Wheres,
And seven million Whys!

Rudyard Kipling

Index of Questions

SECTION 1
Histology and general aspects

1. What is connective tissue?
2. What are fibroblasts and what do they do?
3. What is the structure of a collagen fibre?
4. What is the structure of a reticulin fibre?
5. What is the structure of an elastin fibre?
6. What does the ground substance of ordinary connective tissue consist of?
7. What forms of connective tissue make up the skeleton?
8. What is the structure of tendon?
9. How does tendon insert into bone?
10. How is tendon nourished?
11. How are bones joined together?
12. What are some important characteristics of ligaments?
13. What is the structure of adult hyaline cartilage?
14. How is articular cartilage nourished?
15. How do articular surfaces move on each other?
16. What is fibrocartilage?
17. What is elastic cartilage?
18. What is the structure of synovial membrane?
19. What does synovial membrane do?
20. What are osteoprogenitor cells?
21. What are the features of osteoblasts?
22. What are the features of osteocytes?
23. What are the features of osteoclasts?
24. What are bone lining cells?
25. What is the mineral of bone?
26. How is mineral deposited in bone?
27. What forms of bone are there?
28. What is an osteon?
29. How are bone lamellae arranged other than in osteons?
30. What is periosteum?

Histology and general aspects

In this section the questions are directed towards an understanding of the basic structure of connective tissue in general (1–7), and of those connective tissues in particular, relevant to the skeleton (8–30). Questions of a general nature on bone growth and blood supply are added (31–36).

1. What is connective tissue ?

Connective tissue is one of the four basic tissues of the body, the others being epithelium, muscle and nerve. Whereas in epithelia the cells are closely applied to each other in sheet-like layers, or, if they form glands, as follicles, tubules, acini, clumps or cords, the cells of connective tissue are separated by a matrix of fibres (collagen, elastin, reticulin) and ground substance. The type of cell and the nature and packing of the matrix determines the form of the connective tissue. There are diverse variations. In areolar tissue all the components contribute to a soft supporting material. In adipose and lymphoid tissue the cells predominate. In cartilage, bone, dentine, and the cellular form of cement the ground substance predominates. In fibrous tissue and elastic tissue their respective fibres are the main constituent.

2. What are fibroblasts and what do they do ?

Fibroblasts are cells found in close association with the fibres of connective tissue. They most often appear fusiform in shape with long slender processes that can make contact with the processes of other cells and the fibres of the matrix. They have three main functions —

a) **Synthesis**. Active fibroblasts have the characteristics of protein-secreting cells: euchromatic nuclei; extensive rough endoplasmic reticulum; well-developed Golgi complexes; numerous mitochondria and secretory vesicles. They secrete —

i. the triple helix molecules of procollagen. Outside the cell so-called registration peptides are cleaved from the chains of the procollagen molecules to form tropocollagen molecules, which then polymerise into collagen filaments.

ii. the proteins of elastic fibres. These comprise those that form the electron dense filaments that are associated with elastin fibres and the molecules of tropoelastin which polymerise to form the amorphous glycoprotein elastin.

iii. the proteins and glycosaminoglycans that combine to form the proteoglycans of ground substance.

b) **Degradation**. Fibroblasts ingest fragments of fibrils of collagen and subject them to the digestive enzymes of their lysosomes.

c) **Contraction**. Fibroblasts contain contractile proteins that enable them to move and to shorten the collagen fibrils they secrete in the repair process.

There are probably many different populations of fibroblasts throughout the body with differing characteristics in regard to mobility, capacity to replicate, rate of synthesis, response to glucocorticoids and ovarian and testicular steroids, and life span.

3. What is the structure of a collagen fibre ?

The building block of a collagen fibre is the tropocollagen molecule. Each tropocollagen molecule is a long thin triple helix of polypeptide chains. The differing amino acid sequences of the polypeptide chains give rise to the various types of collagen described. Tropocollagen molecules aggregate to form longitudinal rows called **protofilaments**. About five such rows arrange themselves in parallel to form a **filament**. The molecules of adjacent protofilaments are approximately a quarter staggered to each other. Filaments aggregate to form **fibrils** which can be seen under the electron microscope. At this magnification the periodic crossbanding so typical of collagen, caused by the staggering of the tropocollagen molecules, becomes apparent. Fibrils aggregate to form **fibres,** which can be seen under the light microscope. Fibres lie together to form bundles or **fasciculi**. The fibres of bone, tendon, ligament and fibrocartilage are of type 1 collagen. The fibrils of hyaline cartilage are of type 2 collagen.

4. What is the structure of a reticulin fibre ?

Reticulin fibres are very thin fibres of type 3 collagen seen under the light microscope only after special staining. They branch and unite with each other to form extensive networks, which in the skeleton form a support for the cells and vessels of the marrow. Their coating of glycoprotein is responsible for their staining characteristics.

5. What is the structure of an elastin fibre ?

First formed elastin fibres consist of groups of electron dense filaments of glycoprotein. Amorphous elastin accumulates within the groups of filaments, the filaments taking up a peripheral position. In the mature fibre the filaments are relatively few. The molecules of elastin consist of irregularly coiled chains of polypeptides that can elongate and contract to provide the elastic nature of the fibre. Elastin fibres are usually thin, and branch and join together. They can form thicker fibres, as for instance in the ligamenta flava and elastic cartilage.

6. What does the ground substance of ordinary connective tissue consist of ?

In non-mineralised connective tissue the ground substance is made up of the following components —

    i. **Proteoglycans**. A molecule of a proteoglycan is large and complex with a shape that is likened to the cleaning end of a test-tube brush. **A core protein** represents the stem of the brush. **Glycosaminoglycan** molecules represent the bristles.Glycosaminoglycans are chains of repeating disaccharide units. The number of repeating units per chain varies, but about 20 is common. The number of chains per core protein also varies from 6 to 60. The glycosaminoglycans of the skeleton comprise chondroitin 4 sulphate, chondroitin 6 sulphate, keratin sulphate, dermatan sulphate and hyaluronic acid. Proteoglycan molecules, or units as they are often called, are also linked, by small masses of **link protein**, to long chains of hyaluronic acid. 20–100 proteoglycan units can be linked to one chain of hyaluronic acid, to form a **proteoglycan aggregate**. The whole mass of the aggregate is entangled together to form a shape like a piece of cotton wool. Water is attracted into the volume of each proteoglycan molecule, known as its domain, keeping the bristles of the brush outstretched. Chondroitin sulphate has twice the water binding capacity of keratin sulphate.

    ii. **Glycoproteins**. These are molecules in which the protein component is dominant, in contrast to the proteoglycan molecules in which the carbohydrate component forms most of the structure. They are thought to be important in the binding of the cells of the connective tissue to each other and to the proteoglycans of the ground substance and the collagen of the fibres.

    iii. **Tissue fluid**. A small amount of fluid of similar ionic and diffusible composition to blood plasma with some low molecular weight plasma protein is present in the interstices between the proteoglycan domains, allowing a slow percolation and chemical exchange.

    The concentration and relative composition of these constituents determine the consistency of ground substance. The relative amount of ground substance in connective tissue decreases with age.

7. What forms of connective tissue make up the skeleton ?

The skeleton is made of connective tissue, primarily of fibrous tissue, cartilage and bone. Fibrous tissue may be classed as regular or irregular according to whether the fibres are arranged in parallel bundles or as a feltwork. Regular fibrous tissue is found in tendons, ligaments, joint capsules, aponeuroses, tendon sheaths, septa and retinacula. Irregular fibrous tissue is found in perichondrium and periosteum. If a high proportion of elastin fibres are present the fibrous tissue is described as elastic, as in ligamenta flava and elastic cartilage. Marrow contains a supporting framework of reticulin fibres and is described as a reticular connective tissue.

8. What is the structure of tendon ?

Tendon is made primarily of type 1 collagen fibres. They are arranged in fine to coarse fasciculi. Sparse fibroblasts, called **tenocytes**, lie in longitudinal array in the ground substance between the fasciculi. Each tenocyte passes thin processes, often several times the length of the cell, between the fibres that compose a fasciculus. In cross section the cells appear stellate and their nuclei as thin darkly stained bars. In longitudinal section they appear spindle shaped, with less darkly stained oval nuclei. Tenocytes are more numerous in the proximal part of a tendon than the distal. The fasciculi are separated by thin partitions of loose connective tissue called **endotenon**, which contain such vascular and nervous elements as are present in tendon. This tissue reaches to the periphery of the tendon to form a thin surface layer, called **epitenon**, which may merge with the surrounding loose connective tissue called **paratenon** or be covered by a layer of synovial membrane.

The multiple fascicular construction of tendon allows for —

i. interconnections between them, for example, the intertendinous bands between the long extensor tendons, and the chiasma formation of the flexor digitorum superficialis tendons in their digital fibrous sheaths.

ii. multiple insertions, for example, of abductor pollicis longus and tibialis posterior.

iii. a change of profile in section needed to adapt to certain situations such as, again, of the flexor tendons in the digital fibrous sheaths.

Elastin fibres lie scattered sparsely among the collagen fibres adjacent to tenocytes and their processes. Fasciculi are often arranged in spiral fashion. This arrangement, together with the elastic component of the tendon, allows the storage of some energy when the tendon is strongly stretched, even though, macroscopically, the structure appears virtually inextensible.

9. How does tendon insert into bone ?

At the point of insertion of a tendon into bone its fibres pass through two transitional zones. In the first zone tendon becomes fibrocartilage. Over the course of several hundred microns tenocytes change from thin flat cells to plump and oval ones, and arrange themselves into pairs or rows. They take on the appearance of chondrocytes, lying in lacunae of extracellular matrix separating the collagen fibres. In the second zone, also over the course of several hundred microns, the fibrocartilage becomes mineralised. The transition between the two zones is abrupt, almost perpendicular to the course of the fibres of the tendon. Many of the entrapped cells appear to survive, though others degenerate. The mineralised fibrocartilage merges with lamellar bone.

10. How is tendon nourished ?

Tendon is nourished through blood vessels of arteriole and venule size, and by diffusion from synovial fluid from tendon sheath.

Extrinsic blood vessels come from several sources —

i. Proximally perimyseal vessels continue across the myotendinous junction.

ii. Distally, at the insertion of the tendon, vessels of the periosteum are continuous with those of the adjacent epitenon.

iii. Where a tendon is not surrounded by a synovial sheath it traverses loose connective tissue. The connective tissue continuous with the epitenon is called paratenon. Blood vessels reach the epitenon from adjacent vessels of the paratenon.

iv. Where a tendon is surrounded by a synovial sheath the synovial layer covering the tendon is incorporated into the epitenon. The parietal layer of synovium, separated from the tendon by synovial fluid is called **peritenon**. Vinculae are strands of connective tissue called **mesotenon** which are covered by synovium and extend from the peritenon to the epitenon. They carry vessels to the tendon.

The blood vessels of paratenon and from peritenon are coiled, to allow the displacement of the tendon during joint movements.

Intrinsic blood vessels are fed from the surface vessels of the epitenon. They enter the tendon via the endotenon which separates the fasciculi. They anastomose freely by transverse and oblique branches with longitudinally orientated vessels around each tendon fasciculus to form a uniform inter-fascicular plexus. The density of the vascular plexus can vary, the arrange-ment sometimes being of surgical significance. For example, the palmar aspect of a profundus flexor tendon in its sheath is relatively avascular, most of the vessels being in the dorsal aspect, so that sutures are best confined to the palmar avascular area.

Nutrients can diffuse to the centre of a digital flexor tendon from surroun-ding synovial fluid, thereby suggesting an important role for synovial fluid in the nutrition of a tendon bathed by it.

Lymphatic vessels have been found to accompany blood vessels in tendon.

11. How are bones joined together ?

A simple classification of joints is as follows, in order of increasing complexity —

i. A **synostosis** is a complete bony union between originally separate bone surfaces, the fate, for example, of some cranial sutures.

ii. A **syndesmosis** is where the bone ends are joined by fibrous tissue, again, for example between the cranial bones at sutures, and also at the inferior tibio-fibular joint.

iii. A **synchondrosis** is where the bone ends are joined by cartilage as, for example, between an epiphysis and a diaphysis, or between the basioc-ciput and basisphenoid.

iv. A **symphysis** is where the bone ends are joined by fibrocartilage sandwiched between plates of hyaline cartilage as, for example, between the bodies of the vertebrae at intervertebral discs and at the symphysis pubis.

v. A **synovial** joint is where the bone ends are covered with hyaline articular cartilage (or sometimes fibrous tissue), are enclosed within a synovial capsule and have an outer fibrous capsule which is protected by ligaments.

12. What are some important characteristics of ligaments ?

i. They are localised thickenings of the capsule of a joint, or they may be independent of the capsule, lying inside it like the cruciate ligaments of the knee, or outside it like the lateral collateral ligament of the same joint.

ii. Like tendon, ligaments are formed of closely packed parallel bundles of fibres between which lie rows of inactive looking fibroblasts. In most ligaments the fibres are predominantly of collagen, though there are also variable numbers of elastic fibres arranged web-like between them. In the ligamenta flava, however, many elastic fibres are present, elastin making up 80% of the fibrous material. Designed to resist tension the fibres of a ligament are arranged in the direction of stress. At rest they tend to lie in a rather wavy pattern known as crimp, which disappears as soon as stretch is applied.

iii. The histology of attachment of a ligament to bone is the same as that of a tendon.

iv. As a ligament is stretched the following events occur. Any natural slack of the ligament is immediately taken up. The crimp of the collagen bundles disappears. Any obliquely running fibres are straightened. Water and proteoglycans between the fibres are displaced. The bonds between the fibres are strained. If the ligament fibres are stretched to more than 5% of their length microrupture of the bundles occurs causing a mild sprain. Further stretching, with haemorrhage, but with the macroscopic continuity of the ligament maintained, results in a more severe sprain. Further stretching ruptures the ligament.

v. Ruffini corpuscles, Pacinian corpuscles and numerous free nerve endings serve important proprioceptive and nociceptive functions, carrying the senses of position and movement and initiating reflex protective muscular contractions. Loss of such sensation, following the trauma of sprain, rupture or inappropriate surgical incision can lead to instability of a joint.

vi. Some ligaments are reinforced by tendinous insertion of muscles. One example is the extensions of semimembranosus into the medial and posterior parts of the capsule of the knee joint. Another is the extension of the iliotibial tract from the proximal part of the lateral femoral condyle to the tubercle (Gerdy's) in front of the lateral condyle of the tibia, called the anterolateral femorotibial ligament. Gluteus maximus and tensor fasciae latae tense the tract and its distal extension. Such ligaments are said to be **dynamised**.

vii. The beauty of the complex dynamics of ligaments is only just beginning to be generally appreciated. Maintenance of the different length relationships and the exact attachments of ligaments is essential for the normal range of movement and stability of a joint.

13. What is the structure of adult hyaline cartilage ?

Articular hyaline cartilage is the strong, slippery, durable resilient layer, on average about 2–4 mm thick, which covers the bony surfaces which oppose each other in a synovial joint. It is thickest over the patella and thinnest over the interphalangeal surfaces. It is thicker over the centre of a convex surface and the periphery of a concave surface of a joint than it is over the centre of a concave surface and the periphery of a convex surrace. Scanning electron microscopy reveals a gently undulating surface.

It has three constituents: cells called chondrocytes; collagen fibrils; and matrix, which are arranged in four merging zones. These are, from superficial to deep —

a) Zone I or the **tangential** layer. Its cells are flattened, and therefore oval in perpendicular section. Fine collagen fibrils lie tangentially to the surface, in a tight decussating framework. They pass predominantly in one direction, so that a round needle puncture becomes a longitudinal slit or 'split line'. The most superficial part of this layer consists of collagen fibrils and matrix only.

b) Zone II or the **transitional** layer. Its cells are more rounded, and the collagen fibrils pursue an obliquely decussating course.

c) Zone III or the **radial** layer. Its cells are round and larger, sometimes in isogenous clumps of two to four cells. They are often arranged in perpendicular columns. The collagen fibrils are thicker, and pursue a radial, vertical course. They probably intertwine, increasing resistance to lateral shear stress.

d) Zone IV or the **calcified** layer. Its cells are few, and of these a proportion are degenerate. The matrix is impregnated with calcium salts. A blue line, called the tide mark, is evident on H and E stained preparations of decalcified material, along the interface between Zones III and IV. Traumatic separation of articular cartilage usually occurs in this plane. Bundles of collagen fibrils originate in this layer before passing up to the more superficial zones. The junction between this layer and the underlying bone is irregular, giving a 'jigsaw' appearance. The collagen fibrils do not extend across it. Some vessels in immature and large weight-bearing joints penetrate the subchondral bone plate to enter the calcified zone, but do not extend further.

The **chondrocytes** produce and maintain the matrix. They appear metabolically very active. Cytoplasmic processes are present which can vary from short and stumpy to long and occasionally branched. Necrotic cells can be found in all zones. Mitotic figures are not seen. The cell count of adult femoral condylar cartilage is about 15,000 per cubic mm. The number does not reduce during ageing.

The **collagen** is type 2. In the general matrix there are collagen filaments, thin non-banded fibrils and thicker banded ones. Around the lacunae they form perilacunar baskets of fibrils. Collagen fibrils constitute some 10–15% by weight of articular cartilage.

The **matrix** closest to the cells is known as the pericellular or territorial matrix, and has a slightly different texture to the general intervening matrix. The matrix comprises domains of proteoglycans, enmeshed by and linked to the collagen fibrils, and containing 70–80% by weight of the water of which the cartilage consists. There is some variation in the composition of the proteoglycans in different regions of articular cartilage.

The proportion of chondroitin sulphate is greater in territorial than general matrix, and the proportion of heparin sulphate is greater in the deeper zone.

14. How is articular cartilage nourished ?

Since articular cartilage is avascular it must receive its nutrition through either its superficial surface from synovial fluid, or through its deep aspect in contact with vascular bone. A possible third source of nutrition is from the capillaries of the synovial membrane at the periphery of the articular surface. Diffusion of solutes into cartilage from its synovial surface can readily be demonstrated experimentally, and this route is now considered the main nutritional pathway. Indeed if articular cartilage is deprived of synovial fluid, as for instance after a disarticulation, it does not survive. In immature joints and some large weight-bearing joints (hip, knee, head of first metatarsal) vascular channels perforate subchondral bone to reach into the calcified layer. It is probable that immature joint cartilage receives a significant contribution to its nutrition from these vessels, and possible that the deeper cells of thick cartilage in mature joints are also so sustained. In mature smaller joints, however, the subchondral bone plate and the calcified layer are intact and provide a complete barrier to any subchondral nutritional pathway. That subchondral nutrition is not essential for the survival of cartilage is shown by the growth of cartilage which takes place on the surface of loose bodies within a synovial cavity. Also cartilage survives over areas of subchondral bone infarction. Movement of a joint, intermittently compressing or pumping the undulating articular surfaces, is considered probably to play a significant role in the percolation of nutrients through the matrix.

15. How do articular surfaces move on each other ?

Virtually all articular surfaces are now considered to be ovoid in profile, concave or convex, or a mixture of both. They are truly congruent in any particular joint in one position only, at the end of a range of movement, when all the ligaments are taut. In this, the close-packed position, the joint is at its most stable. In all other positions the articular surfaces make reduced contact with each other, one more curved or round surface contacting a less curved or flatter surface. The two contacting surfaces can then move relatively to each other in three ways —

   i. they can **slide**, when the same area of contact of one surface moves over the other.

   ii. they can **spin**, when the same areas of the two surfaces remain in contact but revolve in relation to each other.

   iii. they can **roll**, when the areas of contact change along chords or arcs on both surfaces.

16. What is fibrocartilage ?

Fibrocartilage is found in articular discs and menisci, the articular surfaces of some joints (temporomandibular, sternoclavicular and acromioclavicular), glenoid and acetabular labra, on bony grooves for tendons and at symphyses. Collagen fibrils, of type 2 rather than type 1 as in hyaline cartilage, are the main component, and are arranged in fibres or in lamellae. Ground substance, which is similar to that of hyaline cartilage, is relatively sparse. The cells resemble chondrocytes more than fibroblasts.

17. What is elastic cartilage ?

Elastic cartilage occurs in the larynx (epiglottis, corniculate and cuneiform cartilages, the apices of the arytenoid cartilages), the external ears and the auditory tubes. Elastin fibres pervade the hyaline ground substance between typical chondrocytes. Elastic cartilage does not calcify.

18. What is the structure of synovial membrane ?

Synovial membrane is a smooth, glistening membrane which lines the inner surface of all parts of a synovial joint except the articular cartilage and any menisci. It forms small villi around the margins of the articular cartilaginous surfaces, and opposite the joint line.

Two layers of the membrane are described —
   a) a lining layer next to the joint cavity, called the intima.
   b) a supportive layer, called the subintima.

a) The **intima** consists of cells, from one to four deep, in a relatively compact layer, set in a matrix. The cells can vary in shape from flat to round or polygonal. There are no junctional complexes between the cells and there is no basal lamina. There may be gaps between the cells, wide enough to expose the intimal matrix to the synovial fluid. The cells may have long processes. There are two types of cells, A and B, though intermediate forms are found. Type A cells have a prominent Golgi apparatus, abundant vesicles and vacuoles, but little rough endoplasmic retinaculum. They show more filopodia, mitochondria, filaments and lysosomes than B cells. They have some features of macrophages but also secrete hyaluronic acid. There is evidence that they are derived from circulating monocytes. Type B cells have little Golgi apparatus, scant vesicles and vacuoles, but abundant rough endoplasmic reticulum. They resemble fibroblasts and secrete a protein rich product. Type B cells predominate in human fetal synovial membranes, but type A cells predominate in adult knees. Apparently intermediate types are called AB cells. It may be that type A and type B cells are not distinct, but are interconvertible, their structure reflecting their function at the time of examination. Capillaries can occasionally reach into the intima to just beneath the free surface, and may be fenestrated over adipose tissue.

b) The **subintima** can vary in structure within a joint, according to the predominant underlying tissue component. It may be adipose over a fat pad, areolar over the capsule, or fibrous over, for example, a cruciate ligament. It is richly supplied with capillaries and lymphatics. Free nerve endings, presumably autonomic, are found only in the adventitia of blood vessels. Pacinian corpuscles are found in the boundary zone between the synovial membrane and capsule. Synovial membrane is not therefore sensitive to an incision.

19. What does synovial membrane do ?

Synovial membrane produces synovial fluid, upon which the nutrition of adult articular cartilage depends; it clears waste material from the joint space, and acts as a packing tissue.

i. Synovial fluid is part ultrafiltrate of blood, and part secretion of the cells of the membrane. Certain structural features of synovial membrane, namely its rich blood supply with sometimes attenuated, fenestrated capillaries, a loose cell layer, and a macromolecular matrix, render it a complex filtration mechanism which regulates the movements of solutes, electrolytes and proteins. Thus although the composition of electrolytes and small molecules in synovial fluid is much the same as in blood, there is no fibrinogen, which has a high molecular weight. Some globulin is present in synovial fluid but is in a lower concentration than albumen, again because of the difference in their molecular weights. Hyaluronic acid is a prominent constituent of synovial fluid, and is responsible for its high viscosity. It is probably secreted by the type A cells. The type B cells probably secrete the small amount of protein in synovial fluid which is firmly bound to hyaluronic acid, and the high molecular weight lubricating glycoproteins, known as LGP–1 and LGP–2.

ii. Within synovial fluid are the products of disrupted cells of synovium, articular cartilage and blood, dispersed collagen fibrils, and various catabolic products of matrix turnover. The half life of the proteoglycans of articular cartilage can vary from 3 weeks to 18 months. Small molecule constituents diffuse back through the membrane to the abundant blood and lymphatic supply. Synovial cells, probably type B, secrete a collagenase, for removal of damaged collagen, and plasminogen activator for removal of any fibrin that is formed as the result of damage to the membrane. Macromolecular constituents are endocytosed by intimal cells either directly, or after sinking into the matrix exposed to the joint space. Although type A cells are said to be most like macrophages, all the intimal cells, in conditions such as haemarthrosis and lipohaemarthrosis, appear to be involved in phagocytosis, as do the subintimal macrophages.

iii. By acting as a packing tissue synovial membrane ensures that articular cartilage is in close contact with it, when not in contact with opposing cartilage. The fluid film separating articular cartilage from synovial membrane is normally less then 50 microns thick.

20. What are osteoprogenitor cells ?

They are cells of mesenchymal origin present in the deepest layer of the periosteum, in endosteum, and in the stroma of marrow, which accompany capillary buds in endochondral ossification. They are small, pale staining and spindle shaped, and can only be distinguished from young fibroblasts by their subsequent transformation into chondroblasts or osteoblasts.

21. What are the features of osteoblasts ?

They are polygonal cells, forming a single, epithelial-like layer on the surface of forming bone. They join with each other, and with neighbouring osteocytes, by cytoplasmic processes at gap junctions. They form and then mineralise osteoid. They have the characteristics of protein-secreting cells, namely a euchromatic, eccentric nucleus, extensive rough endoplasmic retinaculum, a large Golgi apparatus and numerous secretory vesicles. They secrete type 1 collagen, proteoglycans, various collagen and calcium-binding proteins, and form matrix vesicles. Matrix vesicles are tiny membrane bound bodies containing high concentrations of enzymes which are thought to play a major role in crystal nucleation. Osteoblasts arise from osteoprogenitor cells. They do not divide. It is thought they may become osteocytes or bone lining cells or revert to osteoprogenitor cells. They are responsive to local stress; to growth hormone, corticosteroid, T3 and T4, testosterone and oestrogen, and to parathyroid hormone via receptors in their cell membranes. It is thought that osteoblasts respond to parathyroid hormone by liberating a mediator that both reverses the inhibitory effect of calcitonin on osteoclasts, and by directly stimulating the activity of osteoclasts.

22. What are the features of osteocytes ?

They are lens-like cells, spindle shaped or round in section, scattered throughout and between the matrix lamellae of mature bone. Each cell is separated by a variable extracellular fluid-containing space from a very thin layer of osteoid which lines the cavity or lacuna in which it lies. They are derived from entrapped osteoblasts. From each cell up to a hundred thin cytoplasmic processes or dendrites extend through narrow canaliculi to link with the processes of other osteocytes, lining cells or osteoblasts by gap junctions. Mature osteocytes contain only a narrow rim of cytoplasm with fewer organelles than osteoblasts or young osteocytes, but enough to suggest that osteocytes are active in the maintenance of matrix. They are thought to survive about 25 years. They are essential for the maintenance of bone. Their cell membranes possess parathyroid hormone receptors. The response of osteocytes to the hormone is as yet uncertain. It may be to assist in calcium retrieval from the surrounding matrix by stimulating membrane calcium pumps, withdrawing calcium from the surrounding fluid, and passing it, by gap junctions to bone lining cells, and thus on to the bloodstream.

23. What are the features of osteoclasts ?

Osteoclasts are the cells responsible for the breakdown phase of bone remodelling. They are large, irregularly branched, motile cells containing up to 50 or more nuclei, though mononuclear forms can also be found. They are characteristically scattered over resorbing surfaces of bone, the larger cells lying in concavities, called Howship's lacunae, excavated by their enzyme action. They do not divide; their fate is uncertain. The surface of bone against which the osteoclast lies presents the picture of destruction. The bone crystals are dissolved, leaving fragments of collagen fibres exposed. The face of the cell directed to the dissolving bone surface is characterised by an area of ruffled membrane made up of closely spaced, branched, finger-like processes. Surrounding the ruffled membrane is a zone of smooth membrane and clear cytoplasm lying close to the bone. This surface appears to act as a seal to contain the intense activity evident within its perimeter. The cytoplasm deep to the ruffled border is described as the vesicular region of the cell where lysosomes are active in the digestion of the products of endocytosis. Deep to the vesicular is the basal region of the cell, where lie the organelles typical of active cells. Osteoclasts do not have receptors for parathyroid hormone. They respond to the hormone indirectly via a mediator from osteoblasts by increasing in number, and by increasing the size and activity of their ruffled borders. They may be of prime importance in the massive release of calcium from bone which occurs in response to prolonged stimulation by parathyroid hormone, but there is little evidence that osteoclastic destruction of bone has a significant role in normal calcium homeostasis. Calcitonin has the opposite effect of decreasing the size of the ruffled borders, to the point of disappearance, and of slowing the rate of osteoclast formation.

24. What are bone lining cells ?

They are flat cells, described as arranged as a seemingly inactive epithelial layer, on endosteal, periosteal and Haversian canal bone surfaces, where neither bone formation or destruction is occurring. They show gap junctions between themselves and neighbouring osteocytes. Their function and potentialities are uncertain.

25. What is the mineral of bone ?

About one third of bone by weight is organic matrix, mostly type 1 collagen. The remaining two thirds consists of mineral. This mineral is probably first deposited in the form of amorphous calcium phosphate, which is quickly converted to rhombic crystals of calcium hydroxyapatite, growing to a size of about 3 by 25 by 40 nm. They are described as poorly formed and disorganised, however, containing traces of minerals beside calcium, such as sodium, zinc and magnesium, with significant amounts of citrate and carbonate. A hydration shell of ions and water surrounds each crystal, which allows exchange of ions between body fluid and the slightly unstable external and internal structure or the crystal. The relatively large surface area for the size of the crystal assists this process. Fluoride ions can exchange for hydroxyl groups. Lead, radium and strontium among other elements can substitute for calcium.

26. How is mineral deposited in bone ?

Bone crystals form either within or in close association with matrix vesicles. These are minute outgrowths of osteoblasts that round off and separate. They contain enzymes, especially alkaline phosphatase, capable of initiating calcium phosphate deposition, and may act as sites of crystal nucleation. Foci of nucleation may also be formed by modification or degradation of proteoglycans, especially of those containing sulphated glycosaminoglycans. The crystals of hydroxyapatite seem to be deposited first in the gaps between the molecules of collagen on the surface of the fibrils and then alongside them. It may be that collagen also acts as a precipitating mechanism, the galactose groups of collagen molecules forming sites of nucleation.

27. What forms of bone are there ?

On a macroscopic level bone may be —
    i. **Compact**. It is dense, forming the thick walls of the tubular shafts of long bones. It forms the outer shells of flat bones, the ends of long bones and around irregular bones. It forms the underlying support of articular surfaces. The shaft of a long bone is subject to torsional and angular, as well as tensile, compressive and shear stresses. The dense texture of its compact bone, its cylindrical design, and its trabecular organisation are best suited to resist these combined forces.
    ii. **Cancellous**, spongy or trabecular. It is honeycombed with spaces breaking the bone up into a lattice work of connecting struts. It is particularly well developed in long bones towards their thrust-bearing articular surfaces, where torsional, angular and shear stresses are obviated by the virtually friction-free hyaline cartilage covered articular surfaces. Only compressive stresses remain and these are best resisted by the expanded ends of the bones in which the trabeculae are spread out, aligned in the direction of stress and buttressed transversely by other trabeculae passing between them. Patterns of trabeculae, described as trajectories, are characteristic of individual bones and are well demonstrated on X-ray films, on which they should be routinely studied.
    On a microscopic level bone may be —
    i. **Woven**, primary or immature. It characteristically occurs in fetal bone and in the repair of fractures. It is also present in bone tumours and in Paget's disease. It is newly laid down bone. Bone crystals are irregularly deposited among collagen fibrils passing in all directions, so that there is no

evidence of organisation into lamellae. The matrix appears unstriped in stained sections. The cells of woven bone are more closely packed than in mature bone and their lacunae are round instead of elongated. It also contains more proteoglycans than does mature bone.

ii. **Mature**, secondary or lamellar bone. Bone crystals are regularly deposited among fibrils arranged in parallel array, in lamellae formed into osteons. The matrix appears striped in stained sections.

28. What is an osteon ?

An osteon, or Haversian system, is the basic unit of mature compact bone formation. It is a branching, anastomosing and sometimes blind-ending cylinder made up of cell-permeated, concentric, mineralised lamellae surrounding a central neurovascular canal.

Several types are distinguished —

a) a **mature** or secondary osteon. Each consists of —

i. a central canal averaging about 50 microns in diameter. It contains one or two fenestrated capillaries with surrounding pericytes, a number of unmyelinated nerve fibres, bundles of collagen fibrils especially peripherally, and bone lining cells that send numerous processes out into minute bone canaliculi. No lymphatic vessels are found.

ii. up to about 15 concentric mineralised lamellae. The collagen fibrils of adjacent lamellae run in different directions. Lamellae with circumferentially orientated fibrils alternate with lamellae with longitudinal or oblique fibrils. Fibrils within lamellae may also show some degree of variation in orientation. Collagen fibrils of adjacent lamellae intermingle, there being no zone separating them. Osteocytes lie in fluid-filled, lenticular lacunae between the lamellae. They communicate with each other via gap junctions between the numerous long thin processes which radiate through minute canaliculi between the lacunae. Extracellular fluid fills the space between the cell membrane and the crystals of the lacunar and canalicular walls.

iii. a surrounding **cement** or **reversal line**, a layer containing little or no collagen, a high content of proteoglycans and perforated by only occasional canaliculi.

b) a **primary** osteon. Whereas a secondary osteon is formed by the filling in of an absorption channel through pre-existing osteons, a primary osteon is one first formed within the spaces between lamellae of woven bone.

c) an **absorbing** osteon or absorption channel. Osteoclasts line the edge of the canal. Numerous cytoplasmic extensions reach out to the matrix separating bone crystals from collagen fibril remnants. Erosion bays or Howship's lacunae are formed. The central canal enlarges as the lamellae disappear. A resorption canal is relatively short and forms what is called the front part of a cutting cone of a bone-remodelling unit. Behind the cutting cone is found the closing cone of a developing osteon.

d) a **developing** osteon. The central canal of a developing osteon is at first relatively large and is defined by a cement line. Numerous osteoblasts and undifferentiated cells are present in the canal. Concentric lamellae are laid down within the cement line entrapping osteoblasts which then take on the characteristics of osteocytes.

29. How are bone lamellae arranged other than in osteons ?

Bone lamellae may also be arranged —

i. in compact bone as continuous **circumferential** lamellae at the surfaces of the bone, where they are produced by the periosteum, or as fragmentary remains of pre-existing osteons that have been partly removed by absorption channels.

ii. in the trabeculae of cancellous bone lying parallel to the surface, or as angular fragments in a somewhat mosaic pattern. They too are the remains of pre-existing osteons with some endosteal deposition.

30. What is periosteum ?

Periosteum is the fibrous layer covering the non-articular surface of a bone. It is absent at tendon and ligament insertions, and over the surface of sesamoid bones, e.g. the patella. Interestingly, the connective tissue layer covering the subcapsular surface of the neck of the femur in an adult does not seem to have an osteogenic potential and therefore does not perhaps qualify as periosteum. Periosteum is described as made up of two merging layers. An outer layer consists of dense fibrous connective tissue with its fibres generally parallel to the surface. It contains fewer, but larger vessels than the deeper, osteogenic or cambial layer. Muscle fibres take origin from this layer. Vessels of the fibrous layer at these sites link up with those of the muscle, thereby connecting with the muscle 'pump'. The osteogenic layer is of more loosely arranged vascular connective tissue. If this layer is inactive, i.e. not forming bone, the cells are few and are indistinguishable from fibroblasts. If active the osteogenic layer contains many osteoblasts in contact with the bone surface. Periosteum is held to the underlying bone by bundles of collagen fibres, called Sharpey's fibres, which penetrate through the circumferential lamellae. These are seen particularly, for instance, holding the periosteal layer of the dura to the floor of the cranial cavity. Elastin fibres also penetrate into bone. Vascular channels pass in from the periosteum and play an important part in the circulation of blood through the bone. The periosteum of children is relatively thicker, especially in its osteogenic layer, than in adults. It also has greater osteogenic potential and it is more easily elevated from the underlying bone. It is relatively stronger than in an adult, less frequently rupturing all around a fracture. It is, however, strongly attached to the periphery of growth plates. Though bone is insensitive to the penetration of a drill or pin, periosteum has a sensory nerve supply and needs to be anaesthetised before such a procedure can be carried out without a general anaesthetic.

31. How is bone formed in membrane ?

Bones formed in membrane include the frontal, parietal, squamous temporal and the interparietal part of the squamous occipital bones (most of the
vault of the skull), the bones of the face, most of the mandible and the
middle of the clavicle. Subperiosteal bone formation is also a form of
intramembranous bone formation. The 'membrane' in which the new bone
appears in the fetus consists of a layer of vascular mesenchyme, that is cells
derived from embryonic mesoderm lying in a matrix of randomly orientated
fine collagen fibrils in ground substance pervaded by numerous capillaries.
Thin bars of denser matrix or osteoid appear between the cells as the
collagen fibrils enlarge, increase in number, and form into trabeculae. The
mesenchymal cells enlarge, multiply, gather on the surface of the osteoid,
transform into osteoblasts and lay down spicules of bone. The trabeculae
become thicker and longer and join to form an irregular network around
enclosed vessels. The osteoblasts continue to lay down bone on the surface
of the trabeculae, enclosing the spaces between them. These become lined
by concentric lamellae. Osteoblasts entrapped in the osteoid become
osteocytes. The original mesenchyme is transformed into a weave of primary
osteons which soon begin to remodel by absorption and new, secondary
osteon formation. Bones other than those formed in membrane or by
intramembranous ossification, that is the majority, are formed by replacement of a cartilage model by the process of endochondral ossification.

32. How does a long bone form ?

A number of stages are involved —

    i. A **cartilage model** of the bone is first formed. Mesenchymal cells
enlarge, become darker staining, come closer together, transform into
chondroblasts and secrete chondroid, the matrix of cartilage. The bone
thereafter is said to be formed in cartilage, or by endochondral ossification.

    ii. A **centre of ossification** forms in the middle of the shaft or
diaphysis of the cartilage model. The cells in the middle of the cartilage
hypertrophy and die as calcium phosphate crystals are deposited in the
matrix. The central matrix becomes a system of fenestrated septa. A
subperiosteal collar of bone is laid down around the diaphysis by the cells
of the perichondrium transforming into osteoblasts. Blood vessels grow in
from the collar invading and absorbing the central area of empty spaces to
form a marrow cavity. The vessels then grow towards either end of the
cartilage model penetrating into newly formed cavities formed by the further
multiplication, hypertrophy and death of cells between calcified septa. New
bone is laid down on the surfaces of the septa by osteoblasts carried in with
the vessels to form woven bone. Towards the ends of the diaphysis, called
the metaphyses, regular growth of cartilage cells gradually form into growth
plates or physes.

iii. **Secondary centres of ossification** develop usually but not always at both ends of the bone, to form epiphyses, by the ingrowth of vessels into central areas of empty calcified matrix. They are separated from the metaphyses by the growth plates.

iv. Growth in length of the bone occurs at the growth plates, growth in girth by subperiosteal deposition, and remodelling of the bone occurs by a mixture of resorption and deposition.

v. Linear growth is arrested at maturity by the closure of the growth plates; osteoblast activity on the surface also ceases unless stimulated by increased physical exertion.

33. What is the structure of a growth plate ?

By the time the primary centre of ossification has extended about three quarters of the middle length of the cartilaginous model, the fronts of ossification have become organised into the regular structure of a growth plate. The subperiosteal collar of bone advances no further than the zone of hypertrophic cartilage. The growth plate is responsible for the increase in length of the bone, but can also expand in girth with that of the bone as a whole.

The **main structure** of the growth plate consists of an orderly sequence of zones of change, which from the end of the bone towards the centre comprise —

i. a zone of quiescent-looking cartilage cells.

ii. a zone of growth. The cells multiply to form longitudinal columns.

iii. a zone of cell hypertrophy. They become larger, extending the cell columns. It is not certain whether they die at this stage or transform into osteoblasts at least temporally before doing so. The transverse partitions between the enlarged lacunae break down. The vertical interlacing walls become thinner and impregnated with bone crystals.

iv. a zone of bone formation. The spaces are invaded by ingrowth of capillaries and osteoblasts which lay down bone on the surface of the calcified trabeculac to form woven bone.

v. a zone of remodelling. Accompanying osteoclasts set about the removal of the newly formed trabeculae and further osteoblasts lay down more bone until primary osteons are formed.

The **periphery** of the growth plate is the site of its circumferential expansion. It is limited by a collar of fibrous tissue superficial to a wedge-shaped zone of cells continuous with the zone of growth and known as the zone of Ranvier.

Other anatomical aspects of clinical importance in the structure of growth plates are —

i. the hypertrophic zone of the plate is its weakest. Shear stresses are liable to cause separation of an epiphysis from the metaphysis through it, but since the zone of multiplying cells is unaffected, growth of the bone in length is not impaired. Only if the multiplying cells are disturbed by a fracture passing through their zone, or by a crushing injury will longitudinal growth be limited.

ii. if the zone of Ranvier is damaged asymmetrical closure of the plate can occur.

iii. the stability of a plate often appears to be enhanced by the extension of interdigitations of cartilage, called mamillary processes, into the metaphysis, and by contouring of the plate, by small undulations or major convolutions, as for instance at the lower ends of the femur and tibia or the proximal end of the humerus.

iv. the bone of the metaphysis next to the growth plate is thin and relatively porous and prone therefore to buckling on compression.

34. How does an epiphysis grow ?

An epiphysis, or secondary centre of ossification, begins like a primary centre by the vascular invasion of a central area of cartilage that has undergone the process of change to a network of calcified trabeculae. The centre expands radially, the cartilage cells transforming at its periphery. This peripheral zone of cartilage becomes in effect a spherical growth plate in contrast to the discoid shape at the ends of the primary centre of a long bone. The layering of the various zones, however is not so obvious as in the ossification front of a primary centre and growth is not so rapid. As the bony nucleus develops and spreads towards the metaphysis the cells of this aspect of the epiphysis slow their growth until eventually proliferation in the cartilage between the epiphysis and the metaphysis is virtually exclusively on the side of and in the direction of the metaphysis. At the periphery of the epiphysis under a joint surface some of the cells become the finite population of the articular hyaline cartilage. In some way, not understood, they are physiologically different and incapable of forming bone. It is postulated that they are derived from mesenchyme destined to form also synovial membrane and therefore of a different nature to those destined to form bone. Small round bones like those of the carpus and tarsus ossify in the same way.

35. What types of epiphyses are there ?

There are two types of epiphyses —

i. **pressure epiphyses**, primarily responsible for longitudinal growth at the ends of long bones. Their growth plates are orientated more or less transversely to the long axis of the bone, and may be entirely intracapsular (e.g. head of the femur), partially extracapsular (e.g. upper end of humerus), or entirely extracapsular (e.g. lower end of radius).

ii. traction epiphyses or **apophyses**, found at the bases of bony prominences which are pulled on by muscles. They are not found inside joints and do not take weight across them. They have increased numbers of collagen fibres across the growth plate aligned to resist the forces of traction applied to them, and fewer cells in the proliferative layer. They appear to be formed of fibrocartilage rather than hyaline cartilage. Their rate of growth appears much slower than in pressure epiphyses. The centres of ossification of the tibial tuberosity and the lesser trochanter are examples of apophyses. Interestingly, the centre for the calcaneal tuberosity has the structure of a pressure epiphysis rather than an apophysis. It is, however, subject to considerable pressure by the combined action of the triceps surae and the attached plantar structures, and in the heel strike phase of walking.

36. How does blood flow through a bone ?

a) Blood flowing through the diaphysis of a long bone comes from —

   i. a **principal nutrient artery** which penetrates the cortex at a thick fascial attachment, traverses a nutrient canal in a direction away from the faster growing end and enters the medullary cavity, where it divides into ascending and descending branches. The medullary arteries pass to the metaphyses, breaking up into branches in the cancellous bone. Branches from the medullary arteries pass radially to the endosteal surface and enter Volkman's canals to feed the Haversian systems of capillaries. Volkman's canals are distinguished by their more horizontal course and by their lack of the encircling systems of lamellae that typify osteons. **The circulation is centrifugal**, draining out through the surface of the cortex, also in Volkman's canals, to join periosteal veins. From the periosteum veins pass through fascial or muscular plexuses to the general venous circulation. Branches of the medullary arteries also pass centrally to a venous plexus which forms into a vein which retraces the path of the nutrient artery.

   ii. **small periosteal arteries**, which penetrate the cortex, also from fascial attachments. They serve limited areas and depths of the bone. The circulation from their terminations is also thought to be centrifugal.

b) Blood flowing through the metaphyses of a long bone comes from the end branches of the medullary arteries and from the multiple small **metaphyseal arteries** which penetrate the thin bone all round the perimeter and anastomose with them. Again the flow from these is thought to be centrifugal. If the nutrient vessels are damaged by fracture or surgery the metaphyseal circulation becomes important.

c) Blood flowing through an epiphysis is derived from **epiphyseal arteries**. One or more penetrate the cartilage and ramify through cartilage canals. Branches pass peripherally and end in capillary networks from which small veins retrace the path of the arteries. The canals have been shown to supply discrete regions of the epiphysis. They serve also as a route of the vascular invasion that accompanies the formation of the centre of ossification. Vessels reach the epiphyseal side of the growth plate and nourish the resting and multiplying cells.

d) Blood flowing to the hypertrophic gone of a growth plate is derived mostly from the terminal ramifications of the nutrient arteries. The periphery of the plate receives vessels through the fenestrations in the adjacent metaphysis and from the perichondrium of the zone of Ranvier. However, if the nutrient vessels from the diaphysis are damaged, the multiplying cartilage cells are not affected. There is delay in ossification of the calcified cartilage only for 3-4 weeks until the circulation is restored. On the other hand if damage is sustained by the perichondrial vessels the nearby multiplying cells may become ischaemic and die, when the affected part of the plate will cease to grow. Marked angular deformity of the limb may result.

e) Blood flowing into a small or irregular bone does so through several nutrient and periosteal arteries. The nutrient arteries break up into radiating branches in the cancellous bone to reach the endosteal surface and cortical osteons. Again the flow is centrifugal.

The individual bones

SPINE — GENERAL ASPECTS

37. Identify the following parts of vertebrae:

i. of a **thoracic** vertebra: the body, the superior costal facets, the inferior costal facets, the pedicles, the superior articular processes and facets, the inferior articular processes and facets, the transverse processes with costal facets, the laminae, the spinous process, the superior vertebral notches, the inferior vertebral notches, the vertebral foramen.

ii. of a typical (third to sixth) **cervical** vertebra: the neurocentral lips and the parts of a transverse process — the anterior bar, the anterior tubercle, the intertubercular lamella, the posterior tubercle, the posterior bar, and the foramen transversarium.

iii. of a **lumbar** vertebra: the margins of bone derived from the ring apophyses, the mamillary processes (small projections from the posterior edges of the superior articular processes), and the accessory processes (small projections from the medial ends of the posterior surfaces of the transverse processes below and lateral to the mamillary processes).

38. What are the functional components of a typical vertebra ?

A typical vertebra has several functional parts.

i. The **body** is the main load-bearing part of the bone. Its flat upper and lower surfaces provide stable resistance to axial compression, and its internal structure is adapted to this purpose. Together with the intervertebral discs the bodies form the anterior column of the spine.

ii. The **articular facets** on superior and inferior processes serve to guide and restrain the movements of adjacent vertebrae. The direction of the guided movement, and the degree of restraint provided by these joints varies with the level of the spine. In the cervical region the facets are in the same plane, sloping downwards and backwards. They allow anterior and posterior translation in flexion, extension, lateral flexion and rotation movements of the neck, but provide less stability against displacement than do facets at lower levels. The facets of thoracic vertebrae allow some rotation, and provide stability against forward displacement of an upper on a lower bone. The facets of the lumbar vertebrae restrain rotation movements and also provide, the lower ones particularly, stability against forward displacement. The articular surfaces, situated one above the other, are described as forming two posterolateral spinal columns. These columns also participate in some load-bearing, in the cervical region more than the thoracic, and in the thoracic more than the lumbar. Stability of the spine is not completely lost until all three columns are ruptured.

iii. The **peripheral processes of the posterior arch**, namely the transverse, spinous, mamillary and accessory processes, function as levers to which all the muscles that act on vertebrae are attached. The transverse processes act directly on the pedicles. The spinous process acts on the pedicles indirectly through the laminae. In the lumbar region the accessory processes act through the transverse processes, and the mamillary processes through the superior articular processes. Muscle is also directly attached to the laminae. Laminae have in addition a protective function, especially in the thoracic region, where they overlap slightly. The posterior elements also provide purchase for stabilising ligaments, such as the ligamenta flava between laminae, and interspinous and supraspinous ligaments between spines.

iv. The **pedicles** are the stout struts that transmit the powerful leverage exerted by the posterior peripheral elements to the body of the vertebra. Their strong cylindrical form is suited to their resistance to angular and rotational strains.

39. What are the normal curves of the spine ?

There are four sagittal curves —

i. the **cervical** curve, which is convex forwards, extends from the apex of the dens to the second thoracic vertebra.

ii. the **thoracic** curve, which is convex backwards, extends from the second to the twelfth vertebra. The most prominent point of the convexity is usually the tip of the spine of the seventh vertebra.

iii. the **lumbar** curve, which is convex forwards, and extends from the twelfth thoracic vertebra to the sacral promontory.

iv. the **pelvic** curve, which is concave forwards, and extends from the sacral promontory to the tip of the coccyx.

The cervical and lumbar curves, convex forwards, are described as **lordoses**. The thoracic curve, concave forwards is described as a **kyphosis**. The term is not usually applied to the rigid sacrum.

There is often a slight lateral curve in the upper thoracic region. It is usually convex to the right in right handed persons and to the left in left handed persons, and is thought therefore to be correlated with muscular dominance. However, in transposition of the viscera, with an aortic arch directed to the right, the convexity of the curve is to the left, suggesting that the direction of the aortic arch is a causative factor.

40. What are the costal elements of the vertebrae ?

The costal elements of the vertebrae are those parts which are derived from the costal processes. The costal processes extend laterally from embryonic mesenchymal vertebrae between developing muscle plates of the body wall. Transverse processes develop behind the costal processes. In the **thoracic** region the costal processes develop into the ribs and so do not contribute to the definitive vertebrae. In other regions of the spine they are variously absorbed into the vertebrae, though sometimes they form ribs attached to the seventh cervical or first lumbar vertebrae. In a typical **cervical** vertebra the costal element on each side comprises the anterior bar, the anterior tubercle, the intertubercular lamella, the posterior tubercle, and the lateral part of the posterior bar of a transverse process. Only the medial part of the posterior bar is equivalent to a transverse process of a thoracic vertebra. The transverse processes of the atlas and axis are also mostly derived from the costal elements. The tips of their transverse processes represent the posterior tubercles of the lower cervical vertebrae, the anterior tubercles being very small, at the junctions of the fronts of the transverse processes with the bodies. In the **lumbar** region the costal elements are the outer ends and the anterior parts of the inner ends of the transverse processes. In the sacrum the costal elements are the anterior parts of the first three segments of the lateral masses, including the auricular surfaces. There are no costal elements in the lower part of the sacrum and coccyx.

41. How is the body of a vertebra constructed to withstand load?

The mass of the vertebral bodies becomes progressively greater down the spine in response to the increasing load carried by them. Several features of the structure of a vertebral body appear to reflect adaptations to withstand the quite considerable compression load to which it is subject, especially in the lumbar spine. Upper and lower surfaces are flat to provide stability, and are thickened peripherally at the site of ossification of the ring apophyses. Like bones in general it is not solid. If it were it would crack like crystal under constant, but dynamic load. It would also be relatively heavy. Within the peripheral shell of cortical bone the body is reinforced by numerous trabeculae arranged vertically in the direction of compression load. These trabeculae are in turn reinforced by horizontal trabeculae, to prevent their lateral collapse. There is space for fluid between the lattice-work of trabeculae, adding a further element of incompressibility.

42. How does a typical vertebra develop ?

At birth the cartilaginous model of a vertebra has three centres of ossification, which appeared in the 3rd month of fetal life. There is a centre for the body, and one for each side of the arch. The centre for the body is usually single, but in about 5% of fetuses two centres form which rapidly fuse. They may lie lateral to each other, or one dorsal to the other. The centres for the arches fuse behind during the 1st year after birth, and union with the bodies is complete throughout the column by the 6th year. Just after this the ring apophyses appear. They form from scattered foci of ossification all around the peripheral zones of the cartilaginous plates on either side of the ossifying body next to the intervertebral discs, but to a lesser extent posteriorly than laterally and anteriorly. They eventually fuse with the body in the late teens. They incorporate the fibres of adjacent anuli fibrosi of the intervertebral discs and thus fix them into the body of the vertebra, but they do not contribute to the growth in height of the bone. The body of the vertebra grows longitudinally at the upper and lower plates of cartilage which have the same structure as the growth plate of a long bone. Growth eventually slows down and the cartilage plates thin to become the end plates of the discs within the perimeters of the ossified ring apophyses. Secondary centres appear at puberty for the tips of the spines and transverse processes, and the mamillary processes, which all fuse by the 25th year.

43. How are typical vertebrae joined together ?

Adjacent vertebrae articulate by an anterior interbody joint and two posterolateral zygapophyseal joints. The interbody joint comprises the intervertebral disc, which is a symphysis. The zygapophyseal joints are synovial. Their articular surfaces are of hyaline cartilage. The joints are reinforced by ligaments. Anterior and posterior longitudinal ligaments pass between the vertebral bodies. Ligamenta flava pass between laminae, and interspinous and supraspinous ligaments join spinous processes. Weak intertransverse ligaments pass between transverse processes.

44. What are the components of an intervertebral disc ?

Each intervertebral disc is made up of three components merging together without clear demarcation. A soft, central nucleus pulposus has the consistency of toothpaste. A firm peripheral anulus fibrosus is formed by concentric lamellae of collagen fibres passing between the smooth peripheries of the horizontal surfaces of the bodies of adjacent vertebrae formed by the ring apophyses. Upper and lower vertebral plates of cartilage cover the porous bone surfaces enclosed by the attachments of the anulus and separate them from the nucleus pulposus.

45. What is the structure of the young healthy nucleus pulposus ?

The nucleus pulposus consists of matrix and cells. The matrix is made up of about 80% by weight of water held in the domains of free proteoglycans and proteoglycan aggregates. Also present are thin fibrils of type 2 collagen, some non-collagenous protein and some elastic fibres. The cells are chondrocytes and are found mostly near the end plates. The few scattered groups of notochordal cells found in newborn discs disappear in infancy. Towards the end plates the proportions of the constituents change to that of fibrocartilage, and towards the anulus to that of the highly ordered fibrous lamellae. The nucleus pulposus is thus envisaged as being enclosed in a capsule of collagen fibres.

46. What is the structure of the anulus fibrosus ?

The anulus is made up of about a dozen concentrically arranged lamellae of collagen fibres attached to bone in the area of the ring apophyses. The fibres of successive lamellae slope upwards in opposite directions at an angle of about 65° to the vertical. The lamellae are thinnest and most tightly packed posteriorly. The collagen is mostly type 1. Water-containing proteoglycan gel binds the fibres together. Elastic fibres are also present in the lamellae. They pass vertically, obliquely and circularly. Chondrocytes are found in the deeper part of the anulus, and fibroblasts towards the surface.

47. What is the structure of an end-plate of an intervertebral disc ?

Each end-plate is about 1 mm thick in the lumbar region, proportionally less higher in the spine. It covers the porous surface of the body of the vertebra. The cartilage is mostly fibrocartilage in the adult, but in the young hyaline cartilage is in contact with the bone. The fibres in the deeper part of the plate are derived from the inner lamellae of the anulus which turn in to run transversely in its matrix. The pores of the bone on which the plate rests communicate with the marrow cavity, providing a pathway of nutrition to the disc.

48. What are the functions of an intervertebral disc ?

The functions of an intervertebral disc are twofold, namely to withstand load and to allow, within limits, movements between adjacent vertebral bodies.

a) Deformation of the nucleus pulposus under compression is constrained by the disc end-plates and underlying bone axially, and by the tension created in the oblique fibres of the anular lamellae radially. The strength of the normal anulus is such that the end-plates and bone give way before rupture of the circumferential fibrous lamellae occurs. The slight elasticity of the anulus also provides a means of attenuating forces that would otherwise be directed fully to the end-plates. The anulus thus acts as a shock-absorber. The normal anulus is also capable of withstanding load by means of its own bulk and strength. However, without the tensioning effect of the nucleus it distorts on prolonged compression.

b) While permitting some movement between adjacent vertebral bodies the disc acts as a restraining ligament. Axial distraction is resisted by all the fibres of the anulus. Flexion is resisted by tension in the posterior fibres, as the nucleus is displaced posteriorly. Extension is resisted by the anterior fibres in the same way. Translation and rotation movements are resisted by those oblique fibres of alternate lamellae that are directed in the line of tension. Lateral flexion is resisted by those fibres on the side opposite to the direction of movement.

49. What is the nerve supply of an intervertebral disc ?

A variety of non-encapsulated nerve endings, often quite elaborate in structure, have been found throughout the outer half of the thickness of a disc. The greatest number is found in the lateral part of the disc, the least number anteriorly and an intermediate number posteriorly. Some encapsulated nerve endings are more or less confined to the surface of the lateral sides. On each side nerve endings are derived from branches of a sinuvertebral nerve posteriorly, a ventral ramus posterolaterally and a grey sympathetic ramus anteriorly.

50. How is an intervertebral disc nourished ?

The only vessels to enter a disc are small ones confined to the outermost lamellae of the anulus. Although marrow vessels approach the end-plates closely through the pores of the vertebral bodies none penetrate into the nucleus. The disc is in effect avascular. Nutrition of the disc therefore depends on diffusion through the anulus and through the end-plates. Only the central parts of the end-plates appear to be permeable. It is thought that the contribution from both routes is about the same, at least for uncharged molecules. Spinal movements alternately compressing and relaxing the disc may assist the flux of water and nutrients.

51. What changes occur in a mature intervertebral disc as it ages ?

Changes occur in all three components of an adult disc as it ages. In the **nucleus pulposus** there is a reduction in the amount of proteoglycan, down to about half the adult level. The proteoglycans that remain are smaller, with a lesser proportion of chondroitin sulphate. The amount of non-collagen protein increases. The amount of elastin decreases. There is some loss of water binding capacity. The diameter of the collagen fibrils increases, as does the total amount of collagen. The net effect of these changes is that the nucleus becomes drier and stiffer, and less distinct from the anulus.

In the **anulus fibrosus** the diameter of the collagen fibrils decreases, microscopic tears develop in the lamellae, and cracks and cavities may form between them. In the **end-plates** there is some loss or chondrocytes and occlusion of the vascular channels. The plates become thinner and less permeable. **Grosser changes**, such as sequestration of nuclear material through fissures in the anulus are to be regarded as pathological. Loss of disc height is not now considered a normal feature of ageing, but as an expression of pathological changes within the disc.

52. What are the features of the anterior longitudinal ligament ?

The anterior longitudinal ligament extends up from the front of the upper part of the sacrum, anterior to all the bodies of the vertebrae and intervertebral discs to the anterior tubercle of the atlas, and beyond, reinforcing the anterior atlanto-axial membrane, as a median cord to the basal part of the occipital bone. It is a broad, strong band in the lumbar region, narrower in the thoracic region, and broadening again in the cervical region. Its fibres run longitudinally, attaching to the anterior surfaces of the vertebral bodies near their margins. The deepest fibres are the shortest, between adjacent vertebrae. The most superficial fibres can span several interbody joints. Some fibres insert into the anterior concavities of the vertebral bodies, mingling with the periosteum, blood vessels and nerves. There is only a loose attachment to the anuli of the discs. In the neck the vertical fibres of longus colli abut the ligament. In the lumbar region the medial tendinous fibres of origin of the crura arise from the ligament.

53. What are the features of the posterior longitudinal ligament ?

The posterior longitudinal ligament extends up from the back of the body of the first part of the sacrum, posterior to the bodies of all the vertebrae and intervertebral discs, to the body of the axis, from which it extends as the membrana tectoria to just within the anterior margin of the foramen magnum and to the adjacent superior surface of the basilar part of the occipital bone. In the lumbar and lower thoracic regions it is serrated in appearance, while in the upper thoracic and cervical regions it is uniformly broad. The ligament is fairly adherent to the anuli of the discs, where their fibres intermingle. The serrations of the ligament occur opposite the discs where lateral fibres of the ligament curve outwards. As in the anterior ligament, fibres of the posterior ligament are attached to the bodies of the vertebrae close to their margins. They are not attached to the posterior parts of the bodies, however, where vessels pass through large foramina. Again, as in the anterior ligament, deeper fibres span adjacent vertebrae, and more superficial fibres span several vertebrae.

54. What are the features of the ligamenta flava ?

Ligamenta flava are so-called because of their 80% content of elastin. They are present throughout the length of the mobile vertebral column, extending, on each side and at each level, from the uppermost part of the posterior aspect and upper border of the lamina below to the anterior aspect of the lower part of the lamina above. The ligamenta meet in the midline between the spines, but allow passage of communicating veins between them. Laterally they merge with the anterior parts of the capsules of the zygapophyseal joints, virtually replacing them. They are thickest in the lumbar region, and thinnest and broadest in the neck. They help form a smooth posterior surface to the vertebral canal, and provide an elastic resistance to sudden flexion movements which might otherwise cause unwelcome stress on the anterior parts of the bodies as they hinge on the nuclei of the discs. Their elastic nature may also serve in extension to prevent them bulging forward and encroaching into the vertebral canal.

55. What are the features of the interspinous and supraspinous ligaments ?

The interspinous ligaments pass between the spinous processes of mobile vertebrae. They are best developed in the lumbar region, where their fibres pass dorsally and cranially. At each level those of the most anterior part of the ligament arise from the posterior part of the vertebral arch on each side of the back of the site of union of the ligamenta flava, and insert into the anterior part of the inferior border of the spinous process above. Here, only, the ligament is a paired structure, the laminae being separated by a slit-like cavity containing fat. The fibres of the middle part of the ligament arise from the anterior part of the superior border of the spinous process below and insert into the posterior part of the inferior border of the spinous process above. Those fibres which arise from the posterior part of the superior border of the process below pass over the top of the spinous process above to form the supraspinous ligament. These fibres bridge the gap between the tips of the spines. The needle, in lumbar puncture, passes through these ligaments. There is no supraspinous ligament, however, between the first sacral and fifth lumbar spines. The interspinous and supraspinous ligaments of the thoracic region are narrow and elongated. The **ligamentum nuchae**, in the neck, is regarded as homologous with the interspinous and supraspinous ligaments in the other regions. It is a dense bilaminar membrane of fibrous and elastic tissue extending from the external occipital crest, the tubercle of the posterior arch of the atlas, and the bifid spines of the cervical vertebrae, to a free border between the external occipital protuberance and the tip of the spine of the seventh cervical vertebra. It serves as a septum for the neck extensor muscles, and for the attachment of trapezius, splenius and serratus posterior superior. In quadrupeds each lamina acts more like a ligamentum flavum.

56. What are the ranges of movement of the regions of a young adult spine ?

These have been estimated to be as set out in the table below.

	Flexion	Extension	Lateral flexion	Rotation
Cervical	40°	75°	50°	60°
Thoracic	45°	25°	20°	35°
Lumbar	60°	35°	20°	5°
Total	**145°**	**135°**	**90°**	**100°**

THE CERVICAL SPINE

57. What are the features of a typical cervical vertebra ?

i. The **body** is relatively small and broad. The superior surface is slightly saddle-shaped, with a bevelled anterior border and lateral, upwardly projecting uncal processes or neurocentral lips. The inferior surface is reciprocally shaped, with a downwardly projecting anterior border. On each side in front there is an impression for the attachment of fibres of longus colli, and behind there are prominent vascular foramina.

ii. The **pedicles** project dorsolaterally from the middle of the body, bordering vertebral notches of equal depth.

iii. The **laminae** are long and narrow and surround, with the pedicles, a relatively roomy, triangular vertebral canal.

iv. The **spine** has a bifid, often asymmetrical tip.

v. The articular processes form pillars, the parallel round facets sloping downwards and backwards.

vi. The **transverse processes** are characteristic. Each has a posterior bar, a posterior tubercle, an intertubercular lamella, an anterior tubercle, and an anterior bar, bordering a foramen transversarium.

58. What are the features of the atlas ?

The atlas consists of two lateral masses, joined by a short anterior and a longer posterior arch. There is no body. The lateral masses carry superior, variably waisted, concave, slightly inwardly facing facets for articulation with the occipital condyles, and inferior, circular, flat to slightly concave and inwardly facing facets for articulation with the axis. The anterior arch shows a small anterior tubercle and a posterior facet for articulation with the dens. The posterior arch has a small posterior tubercle in place of a spinous process, and a groove, superiorly and laterally, often converted into a tunnel called the foramen arcuale behind each superior facet, for passage of the vertebral artery and suboccipital nerve. The transverse processes are wide, the tips representing the posterior tubercles of the lower processes. On each side the spinal accessory nerve crosses the tip of the transverse process. There is a foramen transversarium in each transverse process.

59. What muscles are attached to the atlas ?

On each side the muscles attached to the atlas are —

Longus colli, its upper oblique part, to the anterior tubercle.

Rectus capitis anterior, to the front of the lateral mass.

Inferior oblique, to the undersurface of the transverse process.

Intertransverse fibres medial to inferior oblique.

Rectus capitis lateralis in front, and **superior oblique** behind, to the upper surface of the transverse process.

Scalenus medius, levator scapulae and splenius cervicis, from in front to behind, to the inferolateral aspect of the outer part of the transverse process.

Rectus capitis posterior minor, to the posterior tubercle.

60. How does the atlas develop ?

At birth there is a separate centre of ossification for each lateral mass. The posterior arch is formed by extension and fusion of bone from the lateral masses by the 3rd year. The anterior arch is usually formed from a separate centre which appears about the end of the 1st year, and unites with the lateral masses about the 7th year. The lines of fusion extend across the anterior parts of the superior articular facets and can be mistaken for fractures. There may be two centres for the anterior arch, or it may form, like the posterior arch, by simple extension from the lateral masses. The atlas is unique in that it has no centrum, which is absorbed into the axis. The anterior arch is formed from a hypochordal bow. This is a span of mesenchyme ventral to the centrum, connecting the two costal processes. It is prominent only in the upper cervical vertebrae. It is only in the atlas that the hypochordal bow chondrifies and then ossifies.

61. What are the features of the axis ?

The dens surmounts the body. It has a facet anteriorly to articulate with a facet on the posterior surface of the anterior arch of the atlas and a groove posteriorly facing and articulating with the transverse ligament of the atlas. The superior articular facets are large and oval, and extend from the body onto the pedicles. They face upwards and laterally. The inferior facets are round and extend between the pedicles and laminae. They face anteroinferiorly in line with the facets below. The articular processes of the axis therefore do not form an articular pillar as they do in the lower typical cervical vertebrae. The transverse processes are small with blunt tips representing the posterior tubercles of the lower processes. The foramina transversaria incline upwards and laterally. The laminae are thick and the spinous process is strong. The vertebral foramen is spacious.

62. How does the axis develop ?

The presence of the dens adds to the complexity of the development of the bone, but knowledge of the sequence of ossification is important for the proper interpretation of X-ray films in cases of injury to the neck in young children. The axis has **four** centres of ossification at birth, one for the body and one for each side of the arch in the normal way, but one also for the dens. The nucleus for the dens develops before birth from two centres which fuse except at the tip where they leave a V-shaped gap. Persistence of a translucent vertical line on the X-ray film at the site of fusion of these two centres can look like a fracture. On the open-mouth, anteroposterior X-ray view, the dens of the infant appears to sit, like a stopper in a bottle, on the body below and between the two arches around its lower part. The epiphyseal lines make an H-shape. These all disappear between the ages of 3 and 6, but again a persisting narrow, sometimes sclerotic transverse line at the base of the dens just below the upper surface of the body may look like an undisplaced fracture. As the epiphyseal lines disappear another centre appears in the tip of the dens which fuses with the rest of the dens by puberty. It may persist as the terminal ossicle of the dens, yet another appearance liable to misinterpretation. The lower ring epiphysis forms in the normal way. The lower part of the dens is interpreted as being derived from the centrum of the atlas. The upper part of the dens is formed from the cranial part of the sclerotome of which the caudal part forms the upper part of the atlas.

63. What muscles are attached to the axis ?

On each side the muscles attached to the axis are —

Longus colli, its vertical part, to the shallow hollow on the anterior surface of the body.

Scalenus medius, levator scapulae and **splenius cervicis**, from in front to behind, to the inferolateral aspect of the outer part of the transverse process.

Intertransverse fibres to the upper and lower aspects of the transverse process, lateral to the foramen.

Rectus capitis posterior major to the crest on the posterosuperior aspect of the spine.

Inferior oblique from the upper part, and **semispinalis cervicis** to the lower part, of the lateral aspect of the spine, extending forwards to the lamina.

Multifidus to the undersurface of the spine and lamina.

64. How is the seventh cervical vertebra identified ?

The spinous process is strong, long, horizontal and not bifid. There may be only a small, sometimes duplicated foramen transversarium. It transmits a vein and a grey ramus to the seventh cervical nerve from the inferior cervical or stellate ganglion, or it may be absent. The foramen transversarium on the left side may give passage to the vertebral artery.

65. What are the joints of Luschka ?

The joints of Luschka, also known as the neurocentral or uncovertebral joints, are clefts which appear during the second decade between the uncal or uncinate processes and the superior adjacent bodies of the third to the seventh cervical vertebrae. An uncal or uncinate process, or uncus, is a prominence of the lateral margin of the upper surface of the body of the vertebra that projects towards the bevelled, slightly concave lateral margin of the lower surface of the body of the vertebra above it. The space between seems to become a joint cavity. The space was thought at first to be a synovial joint, but histological examination failed to demonstrate articular cartilage or synovial membrane. The space was later thought to arise by breakdown of the peripheral lamellae of the anulus fibrosus, but the anulus does not extend to beyond the base of the uncus. The space is now thought to be due to absorption of the connective tissue which hitherto filled the gap between the uncus and the apposing body. The explanation is found in the development of the part. The neural arch of a vertebra which displays uncal processes, unlike other vertebrae, develops slightly cranial to the body. The parts of the arch adjoining the centrum become absorbed into the definitive body of the vertebra. As the superior vertebral notch is formed on each side by excavation of the upper anterior part of the remaining arch, next to the body, the part of the arch which is now part of the body becomes the upwardly projecting uncus. The anulus fibrosus extends between the centra so that its outer fibres insert just medial to the base of the uncus. The interval medial to the uncus, between it and the anulus, is filled with connective tissue. In the second decade this tissue is absorbed and a cleft appears, which is the joint of Luschka. In later life the cleft enlarges, fibrocartilage forms, and osteophytes develop, which by transgressing the intervertebral foramen and irritating the spinal nerve passing through it, can cause the brachial neuralgia often associated with severe cervical spondylosis.

66. What are the features of a cervical intervertebral foramen ?

a) The **boundaries** of the foramen comprise the inferior vertebral notch of the vertebra above, the superior notch of the vertebra below, the zygapophysial joint behind, and the joint of Luschka and posterolateral anulus fibrosus in front. The lumen of the foramen is thus liable to encroachment by osteophytes from the joint of Luschka, bulging of the posterolateral part of the intervertebral disc, and swelling, capsular or bony, or subluxation or dislocation of the zygapophysial joint.

b) **Within** the foramen is the dural root sleeve. Within the dural sleeve is the nerve root complex. The sleeve begins when the dorsal and ventral nerve roots, formed by the union of the rootlets of a spinal segment, approach each other and penetrate the dura opposite the intervertebral foramen. In the neck the dorsal root is three times the thickness of the ventral root, reflecting the major sensory role of the cervical roots. The two roots pierce the dura separately, but close together, so that the funnel-shaped root sleeve is at first divided by a lamina which disappears as the roots coalesce distal to the dorsal root ganglion in the outer part of the foramen. The subarachnoid space around the dorsal root may reach out to the inner pole of the ganglion. The dural sleeve becomes the epineurium as the spinal nerve emerges from the foramen. The thickest spinal nerve in the neck is the eighth. The fibres are arranged at first in a single fasciculus.

c) The dural sleeve occupies about 35–50% of the cross-sectional area of the foramen, usually in a central position. With flexion of the neck the sleeve is drawn inwards and upwards towards the upper margin of the foramen, and returns to its relaxed central position on extension. The sleeve is not attached to the margins of the foramen so that some sliding in and out can occur with movements of the neck and arm. The main anchorage of a spinal nerve is on the gutter of the intertubercular lamella where there is attachment of the epineurium to the periosteum and to the adjoining pre-vertebral fascia and adjacent musculotendinous slips.

d) Also contained in the foramen are a spinal artery dividing into anterior and posterior spinal canal and radicular branches, veins, a recurrent sinuvertebral nerve, lymphatics and loose areolar connective tissue.

67. How is the cranium joined to the cervical spine ?

The occipital bone is joined —
a) to the **atlas** by —
   i. the **atlanto-occipital joints**. The surfaces of the reciprocally curved, centrally constricted articular facets on the lateral masses of the atlas and the occipital condyles are parts of an envelope of a transversely lying ellipsoid. The long axes of the articular surfaces point anteriorly and medially. Each joint is surrounded by a capsule which is thickest posterolaterally. The synovial cavity on each side may communicate with the bursa between the dens and the transverse ligament.
   ii. the **anterior atlanto-occipital membrane**, a strong layer between the anterior arch and the anterior margin of the foramen magnum, reinforced by a midline extension of the anterior longitudinal ligament and merging with the capsules of the atlanto-occipital joints laterally.
   iii. the **posterior atlanto-occipital membrane**, a thinner layer passing between the posterior arch and the posterior margin of the foramen magnum. It is penetrated inferolaterally on each side by the vertebral artery, and some veins and the suboccipital or first cervical nerve. The border of the membrane overlying these structures may ossify to form what has been called the ponticulus posticus, a bridge of bone converting the passage into the foramen arcuale. The posterior atlanto-occipital membrane is pierced in cisternal puncture.
b) to the **axis** by —
   i. the **apical ligament**, from the tip of the dens to the anterior margin of the foramen magnum. It contains traces of notochord.
   ii. the **alar ligaments**, strong cords, one on each side, from the side of the apex of the dens to the medial side of an occipital condyle. The left alar ligament checks rotation of the head to the right. The right ligament checks rotation to the left.
   iii. the **vertical limb of the cruciate ligament**. The upper band passes up from the transverse ligament of the axis, behind the apical ligament, to the anterior margin of the foramen magnum. The inconstant lower band passes down to the back of the body of the axis.

iv. the **membrana tectoria**, the broad and strong upward continuation of the posterior longitudinal ligament behind the cruciate ligament, which attaches just inside the anterior margin of the foramen magnum to the dura on the adjacent upper surface of the bone.

c) the **ligamentum nuchae**, to the spines of the cervical vertebrae from the external occipital crest and external occipital protuberance. It is a fibroelastic membrane homologous with the interspinous and supraspinous ligaments in the lower parts of the spine, but offers little security to the craniovertebral joints.

68. How does the atlas articulate with the axis ?

The atlas articulates with the axis by three joints —

a) the **median atlanto-axial joint**. A pivot is formed by the dens, held in front by a synovial joint against the arch of the atlas, and behind by the transverse ligament of the atlas separated from it by a bursa, which can communicate with a lateral joint. The transverse ligament of the atlas is a strong, thick band, attached on each side to the medial side of the lateral mass of the atlas. It gives off upper and lower longitudinal bands to form the cruciate ligament, the upper band to the anterior margin of the foramen magnum, the lower, not always present, to the back of the body of the axis. It has a thin layer of fibrocartilage anteriorly to bear against the dens.

b) the **two lateral atlanto-axial joints** between the articular facets of the atlas and axis. The capsule of each joint is reinforced by a posteromedial accessory ligament.

Reinforcing ligaments are the anterior atlanto-axial, which is a continuation of the anterior longitudinal ligament, and a posterior atlanto-axial, between the posterior arches, in line with the ligamenta flava.

The stability of the atlanto-axial joint is dependant primarily on the transverse ligament. The alar ligaments provide further stability, the others to a lesser extent. The distance between the anterior surface of the dens and the back of the anterior arch, as seen on a lateral X-ray film, is critical in the diagnosis of subluxation. If, following an injury, this distance is more that 3 mm in an adult, or 5 mm in a child, the transverse ligament has ruptured. If the distance is 10 mm, then all the important ligaments are torn. The transverse ligament divides the vertebral foramen of the atlas into an anterior third, containing the dens, and a posterior two thirds containing the spinomedullary junction, which is liable to fatal injury if the bones displace on each other.

69. What muscles are attached to the anterior tubercles of the cervical vertebrae ?

Longus capitis, longus colli and scalenus anterior arise from the anterior tubercles of the third, fourth, fifth and sixth cervical vertebrae. The atlas, axis and seventh cervical vertebrae do not have noticeable anterior tubercles.

70. What muscles are attached to the posterior tubercles of the cervical vertebrae ?

The ends of the transverse processes of the atlas, axis and the seventh cervical vertebrae correspond to the posterior tubercles of the typical ones. **Scalenus medius** can arise from all the transverse processes, sometimes missing out the atlas and the seventh vertebrae. It is the most anterior muscle to arise from the posterior tubercle of a typical cervical vertebra. **Levator scapulae** arises behind the scalenus medius from the transverse processes of the atlas and axis, and the posterior tubercles of the third and fourth vertebrae. **Scalenus posterior** arises behind scalenus medius from the posterior tubercles of the fourth, fifth and sixth vertebrae. Slips of **splenius cervicis** insert into the inferoposterior parts of the transverse processes of the atlas and axis, and into the posterior tubercle of the third cervical vertebra.

71. What movements occur in the cervical spine ?

At the **atlanto-occipital joint** only a little more than 10° of flexion-extension movement occurs. It has often been said that virtually no rotation occurs at this joint, but CT examination has shown a range of about 5° to be present to either side. Figures for lateral flexion vary. Some observers report a range of about 10°, others about 30–40°.

At the **atlanto-axial joint** there is about 10–15° of flexion-extension, no lateral flexion and about 40–45° of rotation to either side. About 40–50% of rotation of the cervical spine occurs at this joint.

In the **post-axial part** of the cervical spine the maximum degree of flexion-extension at any one level is 15° occurring at the C5/6 and 6/7 levels. The maximum normal translation on routine lateral X-ray films is 3.5 mm. A figure greater than this implies structural damage. The greatest degree of both lateral flexion and rotation at any one level occurs between C3/4 and 4/5 vertebrae, amounting to about 10° to one side. Below the atlas, because of the downward and backward slope of the zygapophysial facets, rotation is accompanied by lateral flexion and lateral flexion is accompanied by rotation.

The **total range** of flexion-extension amounts to about 115° (40° of flexion and 75° of extension), of lateral flexion about 50° and of rotation about 60°.

72. What is the course of the vertebral artery in the cervical spine ?

The vertebral artery, accompanied by a plexus of veins and a sympathetic plexus derived from branches of the inferior or stellate ganglion, enters the foramen transversarium of (usually) the sixth cervical vertebra. It can, especially on the left side, enter the foramen transversarium of the seventh, or occasionally of the fifth or fourth vertebrae. It ascends through consecutive foramina transversaria in front of the spinal nerves, displacing them posteriorly, and lies medial to the intertransverse muscles. Opposite each intervertebral foramen the vertebral artery gives off anterior and posterior spinal canal branches, and a radicular branch to the nerve root, which pass through the foramen. In this part of its course the artery lies lateral to the neurocentral joints. Osteophytic outgrowths from these joints may encroach upon and significantly compress the artery. After passing through the foramen of the axis the vertebral artery courses laterally to the foramen of the transverse process of the atlas. In this part of its course the artery is slightly looped providing some slack that is taken up during atlanto-axial rotation. Even so some stretching of this segment of the artery occurs between, and some kinking at, the two transverse foramina. From the transverse foramen of the atlas the artery passes medially around the posterior border of the lateral mass of the atlas lateral to the suboccipital nerve over the notch of the posterior arch to pierce the posterior atlanto-occipital membrane or pass through the foramen arcuale of the posterior arch to enter the spinal canal. The artery, or a branch, may occasionally pass through the second intervertebral foramen with the second cervical spinal nerve. Just before it enters the spinal canal the artery lies in the floor of the suboccipital triangle, deep to semispinalis capitis, where it gives off branches to the surrounding muscles. The first important branch of the vertebral artery is the **posterior inferior cerebellar** artery, which arises within the posterior cranial fossa, before it joins its fellow to form the basilar artery. The posterior inferior cerebellar artery supplies branches to structures in the medulla that lie lateral to the hypoglossal nucleus, and the oblique ventrolateral course of the nerve root fibres. If the blood supply from the posterior inferior cerebellar artery is compromised the structures affected, and the symptoms resulting, known as the lateral medullary (Wallenberg's) syndrome, will be —

i. the vestibular nuclei, causing vertigo, nausea and nystagmus.

ii. the inferior cerebellar peduncle, causing ataxia of the limbs on the side of the artery involved.

iii. the spinal tract of the trigeminal nerve, causing loss of pain and temperature sensation of the face and loss of the corneal reflex on the same side.

iv. the nucleus ambiguus, causing hoarseness and dysphagia.

v. the anterolateral sensory funiculus, causing loss of pain and temperature sensation of the opposite limbs and half of the trunk.

THE THORACIC SPINE

73. What are the features of a typical thoracic vertebra ?

The body has the shape of a waisted cylinder, except for its posterior surface which is flat vertically and concave transversely. There are costal hemifacets, upper and lower, at the posterolateral angles of the body, the upper in front of the pedicle, the lower in front of the vertebral notch. The pedicles arise high up the body, forming a negligible superior but a large inferior vertebral notch. The laminae are short, thick and broad. The vertebral foramen is relatively small and round. Long transverse processes show terminal anterior articular facets. The articular processes with their facets form joint surfaces on the line of a circle centred forward of the disc space. The spinous process is prominent and slanting.

74. What structures lie closely in front of the bodies of the thoracic vertebrae ?

a) The anterior longitudinal ligament.

b) The **thoracic aorta**. It begins as a continuation of the aortic arch to the left of the lower border of the fourth vertebra and descends towards the midline to reach the front of the twelfth vertebra in the aortic opening of the diaphragm behind the median arcuate ligament. The left sides of the middle four vertebral bodies are sometimes slightly flattened in relation to the aorta. Nine pairs of intercostal arteries are given off from the thoracic aorta to the lower intercostal spaces. The upper right intercostal arteries are the longer; they cross the vertebral bodies behind the thoracic duct, oesophagus and azygos vein. The upper left intercostal arteries have a shorter course across the vertebral bodies behind the hemiazygos and accessory azygos veins.

c) The **azygos veins**. The azygos vein is formed to the right of the twelfth thoracic vertebra variably by the junction of the lumbar azygos vein with a vein formed by the junction of the right ascending lumbar and subcostal veins. It ascends in front of the bodies of the lower thoracic vertebrae adjacent to the right posterolateral aspect of the aorta, before deviating to the right to arch over the right lung hilum. It joins the superior vena cava at the level of the upper border of the fifth vertebra. The hemiazygos (or inferior hemiazygos) vein arises in the same way as the azygos vein, to the left of the twelfth vertebra and passes up to the left of the midline in front of the lower five vertebral bodies. At the eighth vertebra it crosses to the right, behind the aorta, thoracic duct and oesophagus to join the azygos vein. It receives the lower three left posterior intercostal veins. The accessory hemiazygos (or superior hemiazygos) vein descends to the left of the midline in front of the fourth to the seventh vertebral bodies, and crosses, like the hemiazygos vein, in front of the seventh vertebra, to join the azygos vein. It receives the left posterior intercostal veins from the middle four spaces.

d) The **thoracic duct**. It begins to the right of the midline near the lower border of the twelfth thoracic vertebra by the confluence of abdominal lymphatic trunks, or cisterna chyli. It ascends through the aortic aperture and posterior mediastinum between the descending thoracic aorta on the left and the azygos vein on the right, behind the oesophagus, from which it can be separated by a small recess of the right pleural cavity. At the fifth vertebra it passes to the left into the superior mediastinum to occupy the groove between the oesophagus and the vertebral bodies, in contact with the mediastinal pleura above the aortic arch. It passes in front of the posterior intercostal arteries and the terminations of the hemiazygos and accessory azygos veins.

e) The **oesophagus**. It bulges a little to the left of the midline in front of the lower cervical and upper thoracic vertebrae behind the trachea. As it descends at first to the right of the aorta it bulges slightly to the right until it leaves the lower vertebrae to pass in front and to the left of the aorta through its diaphragmatic aperture in the right crus level with the tenth thoracic vertebra. It is separated from the vertebral bodies by right posterior intercostal arteries, thoracic duct, azygos vein, terminal parts of the hemiazygos and accessory azygos veins and the lower thoracic aorta. Below the level of the lung roots the posterior oesophageal plexus is formed by branches of the vagi

f) **Lymph nodes** in pre-aortic and para-aortic groups.

g) **Sympathetic branches**. The lower two or three ganglia of the thoracic sympathetic chains lie against the corresponding vertebral bodies. The medial branches of the chains, including the greater, lesser and least splanchnic nerves lie against the vertebral bodies.

h) **Pleura**. The posterior part of the mediastinal pleura on each side covers the sides of the vertebral bodies, the posterior intercostal vessels and the lower thoracic sympathetic chains and splanchnic nerves. The mediastinal pleura separates these structures from the pleural cavity and lung.

75. How is the blood supply to the spinal cord related to the thoracic vertebrae ?

Through the intervertebral foramina pass spinal branches of the superior and posterior intercostal arteries. Each spinal branch divides into anterior and posterior spinal canal branches which form longitudinal anastomoses with each other and opposite branches within the spinal canal, and a neuromedullary branch. If fully developed each neuromedullary artery divides into anterior and posterior radicular branches to accompany the roots of the spinal nerves to the cord where they anastomose with branches of the posterolateral and anterior spinal arteries. This standard pattern is frequently modified, however, either the neuromedullary artery being insufficiently developed to be a significant supply to the cord, or else it forms only either an anterior (ventral) or posterior (dorsal) radicular branch. The anterior radicular vessels are the fewest but largest, and the posterior radicular vessels are the most numerous but of smaller calibre. A particularly important vessel is the **arteria magna of Adamkiewicz** which is the anterior radicular branch of the neuromedullary artery from a left lower intercostal artery, most often at the ninth or tenth vertebral level. Blockage of a spinal arterial branch of a posterior intercostal artery at this level, especially the arteria magna, by a dissecting aneurysm, trauma or ligation can cause ischaemia of the spinal cord.

76. How are the spinal cord segments related to the thoracic vertebrae ?

In the fetus the spinal cord extends to the level of the second sacral vertebra. Prior to birth the differential growth of the cord and spinal column results in the apparent ascent of the end of cord within the dura to the level of the body of the third lumbar vertebra. The final position of the end of the cord, at the level of the L1-2 disc, is not reached until the age of 20. A consequence of these changes is that spinal cord segments are at an increasingly higher level relative to their vertebral segment as the spine is descended. Between the first and about the fifth thoracic vertebrae the spinal cord segment is at the level of the tip of the spine of the vertebra two segments above (e.g. T6 spinal cord segment is level with the tip of the spine of T4 vertebra). Between the sixth and the tenth vertebrae the spinal cord segment is at the level of the tip of the vertebra three segments above (e.g. T12 spinal segment is level with the tip of the spine of T9 vertebra). The third lumbar spinal segment is level with the tip of the spine of T11 vertebra and the S1 spinal cord segment is level with T12 vertebra.

77. What movements occur between thoracic vertebrae ?

The thoracic part of the vertebral column is relatively immobile, due to several factors. The intervertebral discs are relatively thin. The ribs, costal cartilages and sternum act as restraining outriggers. The ligamenta flava are relatively short between overlapping laminae. The interspinous and supraspinous ligaments are relatively short between long overlapping spinous processes. The articular processes, however, are orientated to allow rotation, the axis of which at each level lies forward of an intervertebral disc. The total range of rotation in the thoracic spine to one side is estimated to be about 35°, which constitutes about 70% of thoracolumbar rotation. There is also about 20° of lateral flexion.

THE LUMBAR SPINE

78. What are the features of a lumbar vertebra ?

It has a relatively large body which is slightly kidney-shaped in horizontal outline, vertically orientated articular processes that oppose rotation and forward translation, mamillary and accessory processes, wide flat laminae, a broad blunt spinous process, and an oval, triangular or trifid vertebral foramen. It does not have either foramina transversarii or costal facets.

79. How can the fifth lumbar vertebra be distinguished ?

It has characteristic transverse processes. The base of each transverse process is massive, extending forwards beyond the pedicle onto the body of the bone. The lower border of the process is horizontal medially, but laterally it angles upwards and slightly backwards to a blunt tip. The transverse process of the fifth lumbar vertebra is particularly strong to take the strain of the attached iliolumbar ligament. The vertebral foramen looks rather trefoil, and viewed from behind the articular processes outline a horizontal rectangle.

80. Identify the following structures on an anteroposterior X-ray film of a lumbar spine and upper sacrum.

Spinous processes. Transverse processes. Pedicles. Spaces occupied by intervertebral discs. Inferior articular processes. Superior articular processes. Zygapophysial joint spaces. Laminae. Sacral promontory. Arcuate lines. Alae of the sacrum. Anterior sacral foramina. The twelfth ribs.

81. Identify the following structures on a lateral X-ray film of a lower lumbar spine and sacrum.

Sacral promontory. Bodies of the fourth and fifth lumbar vertebrae. Spaces occupied by the intervertebral discs. Pedicles. Intervertebral foramina transmitting the fourth and fifth lumbar spinal nerves. Superior and inferior articular processes. Vertebral spines. The remains of a sacral intervertebral disc.

82. What is the purpose of an oblique view X-ray film of the lower lumbar spine ?

This oblique view, taken most commonly of the fourth and fifth vertebrae, is designed to show better the neural arches and their processes nearest the film. At each level the appearance is likened to that of a Scottie dog. The obvious 'eye' is formed by the medial end of the transverse process seen end-on. The vague 'nose' is formed by the pedicle. The 'ear' is formed by the superior articular process, the 'neck' by the part between the articular processes, the 'front leg' by the inferior articular process, the 'body' by the lamina, the 'hind leg' by the opposite inferior articular process. Depending on race and sex, between 2 and 5% of the population have a defect of the pars interarticularis of one of the lower two lumbar vertebrae, known as **spondylolysis**. On the oblique view it appears as if the Scottie dog has a broken neck. It is usually bilateral but may be unilateral. It is not a congenital condition but is thought probably to be acquired during childhood as a stress fracture in the presence of an hereditary predisposition. If the defects are bilateral the body of the affected vertebra, with the anterior part of its arch carrying the superior articular and transverse processes, may separate through the defect and slip forwards from the posterior part of its arch carrying the inferior articular processes. The body of the affected vertebra, with the spinal column above it, slips forward on the body of the vertebra below. The posterior part of the arch remains in place over the vertebra below. The condition is known as spondylolisthesis.

83. On an oblique view X-ray film of the lower lumbar spine identify the following parts of the 'Scottie dog'.

Transverse process. Pedicle. Superior articular process. Inferior articular process. Pars interarticularis. Lamina. Opposite inferior articular process.

84. What is meant by lumbar lordosis ?

The term is derived from the Greek word meaning 'bent backwards'. It refers to the normal forward convexity of the lumbar spine. The sacrum is tilted such that the upper surface of the first segment slopes downwards and forwards. When standing the angle of the slope is about 50° to the transverse plane of the body through the promontory, and at about 45° when the body is supine. To compensate for the slope, to position the first lumbar vertebra in the same coronal plane as the base of the sacrum, the lumbar spine must bend backwards, thereby forming the lordosis.

85. How is lumbar lordosis measured ?

The degree of lordosis can vary quite markedly between individuals. It is best measured on a lateral X-ray film by using standard reference lines drawn along bone borders. A common method is to draw lines backwards along the upper borders of the first sacral and first lumbar bodies. They meet at an angle of 20–60° or more when lying down and about 40–80° when standing up.

86. What factors contribute to lumbar lordosis ?

The tilt of the upper surface of the body of the sacrum is compensated for in the spine above, up to the first lumbar vertebra, by —

    i. the wedge shape of the lumbosacral intervertebral disc. The anterior height of the disc is some 6 mm greater than the posterior. It causes an angle of about 16° between the facing surfaces of the lumbar and sacral vertebral bodies.

    ii. a similar but lesser asymmetry of the body of the fifth lumbar vertebra, the anterior height of which is about 3 mm greater than the posterior.

    iii. the tone of the spinal extensor muscles.

87. How does the lumbar spinal canal vary in shape ?

In the upper part of the lumbar spinal canal it is transversely oval in section. Lower down it becomes rather triangular. The canal at the fifth vertebral level tends to be trefoil, with lateral recesses.

88. What are the contents of the lumbar spinal canal ?

i. **The spinal dura mater**. Often called the dural tube, sheath or sac, it extends the full length of the canal. Leading from the dural sac are the dural root sleeves around the spinal nerve roots. In the upper part of the lumbar canal these leave the dural sac behind the bodies of the related vertebrae, the lower sleeves being at a slightly higher level than the upper ones. The fifth lumbar root sleeve takes origin from the sac opposite the L4/5 disc, and the first sacral sleeve from opposite the L5/S1 disc. The downward slope of origin of the root sleeves from the dural sac becomes more acute from above, the first lumbar sleeve coming away at nearly a right angle, the fifth sleeve forming an angle of about 45°. Variously developed ligamentous fibres pass between the anterior and anterolateral aspects of the dural sac to the posterior longitudinal ligament, especially at the level of the fifth vertebra, and from the proximal sleeves to the posterior longitudinal ligament and periosteum of adjacent pedicles.

ii. **The lower end of the spinal cord**. In the adult the widest diameter of the lumbar enlargement of the spinal cord is usually at the lower border of the twelfth thoracic vertebra. Below this the cord tapers rapidly, as the **conus medullaris**, to form opposite the L1/2 disc a slender, glistening connective tissue thread, the **filum terminale**. The conus contains the lower sacral and coccygeal segments. The filum extends to the back of the first segment of the coccyx. The conus may end occasionally at the level of the intervertebral disc above or below its normal termination. It rises slightly with full flexion of the spine. At birth the conus ends opposite the third lumbar vertebra, and gradually ascends with age as a result of differential growth. Up to the third month of fetal life the spinal cord occupies the whole length of the vertebral canal.

iii. **The 22 pairs of roots** of the lumbar, sacral and coccygeal spinal nerves, that constitute the **cauda equina**. The dorsal and ventral roots of a spinal nerve pierce the dural sac separately, but close together, to enter a root sleeve. The partition in the sleeve fades as the roots unite distal to the dorsal root ganglion, in the intervertebral foramen, to form the spinal nerve.

iv. **Cerebrospinal fluid** within the subarachnoid space or lumbar cistern of the dural sac, surrounding the cauda equina. It is estimated that about 30 ml of the total volume of 140 ml of cerebrospinal fluid is contained in the spinal subarachnoid space, and a large proportion of this is in the lumbar part of the sac. It is in this space, usually at the L4/5 level that lumbar puncture is performed.

v. **Extradural fat**. It is thinly spread in the narrow interval between the dural sac and canal wall, principally in the posterior angle formed by the ligamenta flava, and around the root sleeves.

vi. In the extradural space, anterior and posterior **internal vertebral venous plexuses**, branches of the anterior and posterior **spinal canal arteries** from the lumbar arteries, and branches of the **sinuvertebral nerves**.

89. What is lumbar stenosis ?

Lumbar stenosis refers to abnormal narrowing of the lumbar spinal canal. The condition may be congenital, when superadded pathological changes in its walls in later life will rapidly cause symptoms of neurological involvement. Such changes can include marginal osteophytes of the bodies, disc bulging, zygapophysial joint swelling, thickening and bulging of the ligamenta flava or spondylolisthesis.

90. What are the boundaries of a lumbar intervertebral foramen ?

Above, is the pedicle of the superior vertebra. In front, from above downwards, are the posterior aspect of the lower third of the body of the upper vertebra, the posterolateral aspect of the intervening disc, and the uppermost part of the posterior aspect of the body of the inferior vertebra. Below, is the pedicle of the inferior vertebra. Behind, the ligamenta flavum extends outwards to form the anterior ligament of the capsule of the zygapophysial joint. Above, the joint posteriorly is the lateral border of the lamina of the superior vertebra.

91. What structures pass through a lumbar intervertebral foramen ?

i. **A dural root sleeve**. This passes out towards the upper part of the foramen, from under the upper pedicle to become the epineurium of the spinal nerve. The dural sleeve contains the dorsal root ganglion, and the ventral root, joining to form the very short spinal nerve. Within the sleeve the dorsal and ventral roots are surrounded by arachnoid mater and cerebrospinal fluid. Immediately outside the foramen the spinal nerve divides into ventral and dorsal rami. The L5/Sl foramen is the smallest of the group. In it the fifth lumbar spinal nerve occupies nearly a third of the available space, and is the most vulnerable to stenosis of its foramen. The other root sleeves occupy between about 10 – 20% of the available space. The sleeve is tethered in the foramen by connective tissue, especially posteriorly to the capsule of the zygapophysial joint. Transforaminal 'ligaments' have recently been described. They are narrow, fibrous bands which cross the outer end of the foramen above or below the spinal nerve. Anomalies of nerve roots occasionally occur. A foramen may not contain a root sleeve, or it may contain two. In the latter case the sleeve may be that of the one missing from the foramen above or below, or it may be supernumary.

ii. **A sinuvertebral nerve**. This is a recurrent branch of the ventral ramus which passes to supply the posterior longitudinal ligament, the posterior anulus of the intervertebral disc of its own foramen and the one above, and the vessels of the spinal canal.

iii. **Three branches of a lumbar artery**. These are an anterior spinal canal branch, a posterior spinal canal branch and a radicular branch. The radicular branch enters the spinal nerve and divides into branches that pass into the dorsal and ventral roots, within the root sleeve.

iv. **Communicating veins.** They pass between the anterior and posterior internal vertebral venous plexuses and the ascending lumbar veins.

92. What does a surgeon mean by a lumbar spinal root canal ?

The surgeon is describing the pathway taken by a pair of lumbar nerve roots as they approach and emerge through an intervertebral foramen. The so-called canal begins behind the disc above the vertebra which forms the upper part of the foramen. At this point the pair of roots may still be within the dural tube. The canal is here medial to the intervertebral foramen above the one from which the nerve emerges, and is in front of the superior articular process of the vertebra below. The middle part of the canal lies medial to the pedicle below which the nerve emerges. The root sleeve commonly begins at this point. In front of this part of the canal is the body of the vertebra giving rise to the pedicle, and behind it is the lamina of the same vertebra, and the superior articular process of the vertebra forming the lower part of the foramen. This part of the canal is called the lateral recess of the spinal canal. The last part of the canal is the intervertebral foramen itself.

93. What are the features of the lumbar zygapophysial joints ?

i. The lumbar zygapophysial **joint surfaces** are vertical, the plane of the joint varying in transverse section from flat to corved C or J shapes. The flat surfaces can be orientated from towards the coronal to towards the sagittal plane. The curved joints are concave posteromedially. In J-shaped joints the wider part may be either in the coronal or sagittal plane. The joints of the two sides at the same level may not be quite symmetrical. In general the joints of the lower lumbar spine have surfaces directed more to the coronal plane than the sagittal, to oppose forward translation as well as rotation of the upper vertebra on the lower, while the joints in the upper lumbar spine have surfaces directed more to the sagittal plane, to oppose rotation rather than forward translation.

ii. The **joint capsules** consist of lateral extensions of the ligamenta flava anteriorly, are thick posteriorly where they are reinforced by fibres of multifidus, and are loose at the upper and lower poles. The capsules have foramina at their poles to allow flow of fat between intra- and extra-capsular spaces during flexion and extension movements.

iii. **Intra-articular structures,** such as connective tissue rims, adipose tissue pads and fibro-adipose meniscoids are described, and may all be sources of internal derangement.

iv. They do play a part in **weight-bearing.** Prolonged standing with a lordotic spine causes impaction of the inferior articular process of a vertebra above into the inferomedial part of the superior articular process and into the lamina of the vertebra below. It is estimated that up to about 19% of the load between the lower vertebrae can be borne in this way by the zygapophysial joints, the load increasing to 70% with pathological disc narrowing.

94. What is the nerve supply of a lumbar zygapophysial joint ?

A great deal of lumbar back pain is thought to originate from degenerative changes in the zygapophysial joints. A knowledge of the nerve supply of these joints is therefore important to the clinician. Each spinal nerve, formed by the union of ventral and dorsal roots, is short, perhaps 1–2 mm only; it divides immediately outside the intervertebral foramen into the larger ventral ramus, which appears to be a direct continuation, and the smaller dorsal ramus, which leaves the spinal nerve at a right angle, and passes backwards towards the upper border of the transverse process of the vertebra below. The fifth dorsal ramus runs back over the ala of the sacrum. Each of the upper four lumbar dorsal rami divide into lateral, intermediate and medial branches. The fifth lumbar nerve divides into intermediate and medial branches only. The medial branch of each dorsal ramus curves round the root of the superior articular process, under the mamillo-accessory ligament, onto the back of the lamina and gives branches to the zygapophysial joints above and below. It also gives branches to the fibres of multifidus connected to the same vertebra and an interspinous ligament above it. Each zygapophysial joint receives innervation, therefore, from two spinal nerves, the one above and the one below. The nerves can be blocked by local anaesthetic around the roots of the superior articular processes, a very useful treatment in some cases of chronic back pain from osteoarthritis of the zygapophysial joints.

95. What movements take place between lumbar vertebrae ?

These are —

i. **Axial compression**. The prolonged pressure sustained by intervertebral discs in the erect posture in a young adult throughout a day can lead to 1–2% loss of body height, amounting to up to a couple of centimetres. This is brought about by a 10% loss of disc height by loss of fluid.

ii. **Axial distraction**. Distraction of a fresh cadaver lumbar spine by a 9 kg load over the course of an hour has been found to separate adjacent vertebral bodies by less than 1 mm, which reverted to about 0.1 mm within half an hour of the release of the weight.

iii. **Flexion**. This movement straightens the lumbar lordosis, only rarely slightly reversing it. In tilting forward on its intervertebral disc the inferior articular process rises up 2–3 mm on the opposing superior articular process. In young adults there is no more than about 10° of flexion between any two vertebrae.

iv. **Extension**. Backward tilting is limited by impaction of the articular processes, and is restricted to no more than about 5° between adjacent vertebrae.

v. **Axial rotation**. This movement is limited by the curvature of the articular facets of the zygapophysial joints, especially between the upper lumbar vertebrae. The joints impact on the side opposite to that to which the body is turning. No more than about 3° or so of rotation are thought possible between two adjacent lumbar vertebrae without some damage occurring to the joints.

vi. **Lateral flexion**. This is a complicated movement which involves rotation as well as tilting. No more than about 5° of lateral flexion occurs at any one level.

vii. **Forward, lateral** or **backward translation** movements, in the horizontal plane, only ever occur significantly under pathological conditions.

96. What are the close anterior relations of the lower two intervertebral discs ?

i. In front of the **L4/5 disc**, from left to right, are the left common iliac artery, the proximal part of the median sacral artery, the right common iliac artery, and the inferior vena cava. The aorta lies to the left of the midline, and usually bifurcates at the lower border of the fourth vertebra.

ii. In front of the **L5/Sl disc**, from left to right, are the left common iliac artery, the left common iliac vein, the median sacral artery, the right common iliac artery and the right common iliac vein. The inferior vena cava is formed by union of the common iliac veins in front and to the right of the body of the fifth lumbar vertebra. The inferior mesenteric artery, as it becomes the superior rectal artery, may lie to the left of the disc, to the right of the bifurcation of the left common iliac artery.

All these vessels are at risk when the discs are operated on from behind if grabbing forceps are pushed too far anteriorly when evacuating central disc material. Injury to the vessels caused in this way has resulted in immediate and sometimes fatal haemorrhage, or later arteriovenous fistulae.

iii. Common to **both discs** is the superior hypogastric plexus of sympathetic nerves. It may be a band of fibres or a more plexiform arrangement. A ureter may lie more medially than usual. The appendix may lie near the right side of the L5/Sl disc and may have been amputated during disc surgery!

THE SACRUM AND COCCYX

97. Identify the following parts of the sacrum and coccyx.

The **promontory**, the projecting upper anterior border of the first sacral vertebra. It forms the posterior median part of the brim of the pelvis.

The **transverse ridges**, the sites of fusion of the bodies of the sacral vertebrae.

The **lateral masses of the sacrum**, formed by fusion of the transverse and costal elements of the sacral vertebrae, which expand above as the alae.

The **four anterior sacral foramina**, communicating through intervertebral foramina with the sacral canal, and transmitting the sacral or pelvic, ventral or anterior primary rami.

The **spinous tubercles**, joined to form the median sacral crest.

The **articular tubercles**, joined to form the intermediate sacral crests, representing fused articular processes.

The **transverse tubercles**, joined to form the lateral sacral crests, representing fused transverse processes.

The **four posterior sacral foramina**, communicating through intervertebral foramina with the sacral canal and transmitting the sacral dorsal or posterior primary rami. They open between the transverse elements of the lateral masses.

The **cornua**, formed by the articular processes of the fifth sacral vertebra, for articulation with the cornua of the coccyx.

The **hiatus**, formed by the failure, normally, of the fifth dorsal arch to form.

The **auricular surfaces**, so-called on account of their shape, for articulation with the hip bones at the sacroiliac joints.

The **areas for attachment of the interosseous ligaments** of the sacroiliac joints.

The **cornua of the coccyx**, representing the pedicles and superior articular processes of the first coccygeal vertebra.

The rudimentary **transverse processes of the first coccygeal vertebra**.

98. What is the anatomical position of the sacrum ?

The upper surface of the body of the first sacral vertebra slopes such that it faces forwards as well as upwards at an angle of between 45–50° to a transverse plane through the promontory.

99. What prevents the fifth lumbar vertebra from slipping forwards on the sacrum ?

Forward slip, or spondylolisthesis of the fifth lumbar vertebra on the sacrum is normally prevented by —

a) The alignment of the **vertical facets of the zygapophysial joints**. At the lumbosacral level the articular facets tend to lie in a more coronal plane than they do higher up the lumbar spine, where they become more sagittal. The lumbosacral facets are aligned to prevent forward slip more than rotation of the lowest lumbar vertebra on the sacrum. The facets at higher levels of the lumbar spine serve to counteract rotation more than forward slip of adjacent vertebra.

b) The **interbody joint**. The bodies of the fifth lumbar and first sacral vertebrae are held firmly by a strong intervertebral disc, reinforced by anterior and posterior longitudinal ligaments.

c) The **iliolumbar ligaments**. These are strong ligaments passing, on each side, from the whole length and tip of the transverse process of the fifth lumbar vertebra to the transverse tubercle of the first sacral vertebra, to the anteromedial aspect of the posterior segment of the crest of the ilium, and to the posterior part of the inner lip of the iliac crest. The upper part of the ligament forms anterior and posterior thickenings of the fascia of the lower attachment of quadratus lumborum. Other fibres pass down anteriorly across the sacroiliac joint to the posterosuperior part of the iliac fossa and posterior extremity of the arcuate line of the hip bone.

There is no supraspinous ligament at the lumbosacral level. The interspinous ligament and the more elastic ligamenta flava are relatively less effective than the other supports in resistance to prolonged strain.

100. What are the anterior relations of the sacrum ?

The anterior relations of the sacrum are —

a) **piriformis**, which arises on each side from the anterior aspect of the bars of bone between the anterior sacral foramina, and the anterior aspect of the lateral part of the bone, from the uppermost to the lowermost foramen.

b) **nerves** —

i. on each side the **lumbosacral trunk**, formed by the anterior primary ramus of the fifth lumbar nerve, joined by a branch of the fourth, passes across the anterior part of the ala medial to the iliolumbar vessels to join the anterior primary ramus of the first sacral nerve on the anterior surface of the piriformis.

ii. on each side the **obturator nerve** crosses the anterior part of the ala lateral to the iliolumbar vessels.

iii. on each side the **anterior primary rami** of the upper four sacral nerves pass out of the anterior sacral foramina onto the anterior surface of the piriformis to form the sacral plexus.

iv. on each side a **sympathetic trunk** passes down medial to the anterior sacral foramina to meet its fellow in front of the coccyx to form the **ganglion impar**.

v. the **superior hypogastric plexus** passes down in front of the promontory, dividing into the inferior hypogastric (pelvic) plexuses.

c) **Vessels** —

i. the **median sacral vessels** pass down in the midline from above the promontory to the lower end of the bone.

ii. on each side the **iliolumbar vessels** pass upwards across the ala between the obturator nerve laterally and the lumbosacral trunk medially.

iii. the **superior rectal vessels** run obliquely in the right limb of the sigmoid mesocolon, from opposite the left sacroiliac joint at the brim of the pelvis, to the front of the third sacral segment, where they divide to supply the rectum. They are accompanied by lymph vessels and a few nodes.

iv. on each side the **lateral sacral vessels** pass down lateral to the anterior sacral foramina, sending branches through them.

d) **Fascia**. The parietal pelvic fascia is relatively thin over the front of the upper two sacral vertebrae, but is strong over the piriformis where it is firmly attached to the medial margins of the foramina. The sacral nerves lie outside the fascia, the vessels within it. The fascia connecting the rectum to the lower part of the sacrum is a thick white layer known as the fascia of Waldeyer.

e) **Peritoneum**. The peritoneum covers the anterior surface of the upper two sacral vertebrae and is reflected forwards to form the right limb of the sigmoid mesocolon. On each side of the rectum the peritoneum forms a pararectal fossa. The sacrogenital folds extend from the sides of the bladder to the sacrum in the male. In the female the corresponding peritoneal folds extend from the sides of the lower uterus and are termed the recto-uterine folds.

f) **Rectum**. The lower three sacral vertebrae are in relation to the fascia of Waldeyer, and to the rectum and its ampulla.

101. What are the contents of the sacral canal ?

The dural sac, with its content of CSF, sacral and coccygeal nerve roots and filum terminale, extends down to the level of the lower border of the second sacral vertebra. Fibrous strands anchor it to the posterior wall of the canal. The filum terminale continues from the apex of the dural sac to the back of the first part of the coccyx. The sacral and coccygeal nerve root sleeves on each side leave the dural sac and pass towards the sacral foramina and sacral hiatus. The roots unite within their sleeves in the canal, which therefore also contain the posterior root ganglia. The upper four sacral nerves divide within the canal and their anterior and posterior rami leave through their corresponding foramina. The fifth sacral and coccygeal nerves, and the filum terminale pass out through the sacral hiatus. Epidural fat in the canal is loose and wide meshed in childhood, but in the adult there is hardly any fat behind the dural sac. Any fat that is present is firm and fibrous, and is more plentiful between the sacral roots. It may cause loculation of injected epidural anaesthetic solution leading to limited and unpredictable neurological effects. An internal vertebral venous plexus, unified below the dural sac, has many communicating veins passing through the sacral foramina and hiatus. These veins communicate with the venous plexuses around the bladder and prostate, and provide a path for the spread of metastases from pelvic growths (particularly from the prostate) to the bodies of the vertebrae. The internal vertebral venous plexus in the sacrum is continuous above with the valveless extradural venous plexuses (Batson's veins) that extend the length of the column to the foramen magnum, and which link with dural sinuses through it. Branches of the lateral sacral arteries pass into the canal through the anterior foramina, supply the contents of the canal, and pass out through the posterior foramina into the erector spinae.

102. What would be the effect of damage to the anterior primary ramus of the first sacral nerve ?

There would be —

i. **Motor loss.** S1 fibres reach to the glutei, hamstrings, muscles of the calf, peronei, and extensor digitorum brevis. Clinically it is most apparent as some weakness of plantar flexion and eversion of the foot, and wasting of the buttock and calf. There may be some weakness of flexion of the knee and wasting of the extensor digitorum brevis.

ii. **Sensory loss.** The S1 dermatome consists of the lateral aspect of the lower leg and ankle, the heel, the lateral part of the foot including the sole and dorsum, and the little toe. It may include also the third and fourth toes.

iii. **Reflex loss.** There will be loss of the ankle jerk.

103. How is the erector spinae attached to the sacrum ?

The bulk of the muscle over the back of the sacrum belongs to **multifidus**, the fascicles of which have arisen from the spines of the lower three lumbar vertebrae. The fascicles arising from the spines of the third and fourth vertebrae insert into the dorsal surface of the sacrum between the transverse and articular tubercles and cover the foramina. The fascicles arising from the fifth lumbar vertebra insert into the sacral groove between the articular and spinous tubercles. The **aponeurosis of the erector spinae** is a strong broad sheet covering multifidus. It is the aponeurosis of insertion of the fibres of the thoracic iliocostalis and longissimus into a U-shaped attachment to the median sacral crest, formed by the spinous tubercles, and the lateral sacral crest, formed by the transverse tubercles.

104. What pass through the posterior sacral foramina ?

Passing through the posterior sacral foramina are —

i. the posterior primary rami of the sacral nerves. The upper three divide into medial and lateral branches. The medial supply multifidus. The lateral communicate with each other, pierce the aponeurosis of the erector spinae and gluteus maximus to reach the skin of the posterior buttock. The lower two do not divide, and supply the skin over the coccyx.

ii. branches of the lateral sacral vessels.

105. What common variations of structure are there of the sacrum ?

i. There may be **sacralisation** of the lowest lumbar vertebra. In this variation the lowest lumbar vertebra is fully or partially incorporated into the sacrum. The angle of 'take-off' of the lumbar spine from the sacrum in such cases may be a little more acute than normal, and it is held by some that an increased risk of the development of degenerative changes in the lumbosacral joints may result. There may be fusion or a pseudarthrosis between the transverse process of the fifth lumbar vertebra with the ala of the sacrum or ilium, which can lead to troublesome symptoms.

ii. There may be **lumbarisation** of the first sacral vertebra. The first sacral vertebra may show a varying degree of independence from the remainder of the bone.

iii. There may be **fusion of the first coccygeal segment** with the sacrum.

v. There may be **spina bifida occulta**. In 10% of the population there is failure of fusion of the dorsal arches of the sacrum not accompanied by any external manifestation of abnormality. It may occur in the lower part of the sacrum, enlarging the hiatus upwards. Or it may occur in the upper part of the sacrum, and be accompanied by similar changes in the lower lumbar vertebrae. The defective neural arches may be of cartilage or fibrous tissue.

v. There may be minor degrees of **agenesis**, as shown by an absent lower part of the coccyx.

106. What uncommon, but more significant, variations of structure are there of the sacrum ?

i. There may be several variations of structure of interest to an anaesthetist or surgeon performing a caudal injection. There may be a fifth posterior arch **narrowing the patency of the sacral hiatus**. The hiatus may be asymmetric or reduced to as little as 2 mm in diameter. There may be quite marked variation in **the volume of the sacral canal**, from 12 to 65 ml (average 33 ml). The anteroposterior diameter of the canal can be reduced to as little as 2 mm and may be curved more than normal. All these factors make it more difficult to pass a straight needle into the canal or judge the correct volume of agent required. The anterior sacral foramina may be larger than normal allowing easier flow of solution into the pelvis.

ii. There may be **hypoplasia of the superior articular facets** leading to one form of spondylolisthesis of the lowest lumbar vertebra on the sacrum.

iii. There may be more **extensive failure of development of the neural arches**. They may be absent altogether in the sacrum. The abnormality may also involve the lower lumbar vertebrae and be accompanied, in 2 per 1000 births, by external manifestations such as a collection of hairs, or dermoid cyst formation. There may be meningocele, meningomyelocele or even myelocele. Neurological deficit can vary from slight to paraplegia.

iv. More **advanced forms of agenesis** can culminate in total absence of the sacrum, with varying degrees of neurological deficit.

v. A **remnant of notochord** is thought to be responsible for the development of the rare chordoma, a tumour which can also occur in the spheno-occipital region. Local bone destruction and neurological damage occurs.

107. What is the structure of the sacroiliac joint ?

    i. The **bony articular surfaces** of the sacrum and hip bone show cranial and caudal limbs placed at an angle to each other, to resemble vaguely an ear. They are thus described as the auricular surfaces of the bones. In the infant they are flat and covered by hyaline cartilage, and are separated by a synovial lined cavity. In the adult, especially in the male, they become irregular, reciprocal elevations and depressions interlocking, supposedly to contribute to the stability of the joint. In later life fibrous adhesions gradually obliterate the joint to a varying degree, sometimes completely. Occasionally these adhesions ossify, thereby abolishing what little movement occurred in the joint.

    ii. The **anterior sacroiliac ligament** is relatively thin except over the arcuate line and along the pre-auricular sulcus below the inferior margin of the caudal limb of the auricular surface of the hip bone.

    iii. The **interosseous sacroiliac ligament** is probably the strongest ligament in the body, attached to the medial surface of the tuberosity of the ilium, and to the rough area behind the auricular surface of the sacrum.

    iv. The **posterior sacroiliac ligament**, again relatively thin, is made up of short upper fibres and longer lower fibres arranged increasingly obliquely from above downwards. They pass from behind the attachment of the interosseous ligament to the hip bone, to the transverse tubercles of the sacrum. It merges laterally with the sacrotuberous ligament, and medially with the aponeurosis of the erector spinae.

108. What movements occur at the sacroiliac joint ?

Before fibrous adhesions within the joint become strong enough to prevent it, a small range of movement at the sacroiliac joint is possible. Because the range is so small, however, it is difficult to analyse and its nature is uncertain. Most commonly described is a rotation of the sacrum about a transverse axis, such that the promontory is displaced a few millimetres forwards, and the apex backwards, the iliac wings medially and the ischial tuberosities laterally. The movements about this axis are called **nutation** and **counternutation**, from the Latin word meaning to nod. In nutation the backward movement of the apex of the sacrum is resisted by the sacrotuberous and sacrospinous ligaments. The transverse axis is difficult to place, and is variously held to be posterior to the auricular surfaces in the interosseous ligament, or in the middle of the auricular surfaces between the cranial and caudal segments, or anterior and inferior to the auricular surfaces, in the pelvis. Also described is a slight linear displacement of the sacrum along the caudal limbs of the auricular surfaces. The only certainty seems to be that the range of movement is more apparent towards the end of pregnancy and during labour, because of the effect of relaxin on the ligaments.

THE STERNUM

109. What are the parts of the sternum ?

The parts of the sternum are —
   i. the manubrium.
   ii. the body or mesosternum, formed of four fused segments or sterne-brae, demarcated anteriorly by transverse ridges.
   iii. the xiphoid process or xiphisternum.

110. What muscles are attached to the manubrium ?

Pectoralis major arises by muscle fibres from the shallow depression which covers the anterior surface of the manubrium lateral to the vertical midline ridge and below the horizontal crest between the lower parts of the clavicular facets.

Sternocleidomastoid arises by its medial, tendinous head, from the lateral part of the horizontal crest.

Sternohyoid arises by muscle fibres from the posterior aspect of the bone, from just medial to the clavicular facet.

Sternothyroid arises with its fellow from a continuous strip of muscle fibres, concave upwards, across the back of the lower part of the bone between the facets for the first costal cartilages.

111. What are attached to the margins of the manubrium ?

Above, in the jugular or suprasternal notch, the fibres of the interclavicular ligament attach to the bone. This ligament forms the lower edge of the suprasternal space of Burns, between the sternal heads of the sternoclei-domastoids laterally, and the two layers of the deep investing layer of cervical fascia attached one in front and one behind to the margins of the manubrium. The anterior jugular veins pass laterally through the space, behind the sternocleidomastoid muscles. **On each side** the clavicle artic-ulates with the facet at the superolateral angle of the bone, but is separated from it by the articular disc of the sternoclavicular joint. This disc is attached at the junction of the clavicular facet and the first costal cartilage. The capsule of the sternoclavicular joint is attached to the margins of the clavicular facet, enclosing the articular disc. Below the facet for the clavicle the first costal cartilage joins directly to the lateral margin of the manubri-um, forming a primary cartilaginous joint, or synchondrosis. Below the facet for the first costal cartilage the lateral margin gives attachment to the anterior intercostal membrane anteriorly and the internal intercostal muscle posteriorly, down to the inferolateral angle of the manubrium. This angle is covered by the articular cartilage of the second, synovial, sternocostal joint. The intra-articular ligament of this joint is attached to the lateral edge of the fibrocartilage of the manubriosternal joint.

112. What lies behind the manubrium ?

The manubrium forms the anterior wall of the superior mediastinum. Between the bone and the **sternothyroid** muscles, branches of the internal thoracic vessels penetrate the foramina scattered over the posterior surface of the bone. The lines of the costomediastinal pleural reflexion pass downwards and medially from behind the sternoclavicular joints to the midline at the manubriosternal joint. Each costomediastinal recess here contains the anterior border of the **lung,** Behind the lower half of the manubrium the **arch of the aorta** passes backwards from right to left, from behind the right half of the manubriosternal joint, up to the level of the middle of the bone, and then down to lie against the lower part of the left side of the body of the fourth thoracic vertebra, level with the left half of the manubriosternal joint. From the arch its three major branches, the **brachiocephalic, left common carotid** and **left subclavian arteries,** pass upwards. The **left brachiocephalic vein** passes downwards and to the right in front of the beginning of the three major arteries from behind the left sternoclavicular joint, to join the right brachiocephalic vein behind the second right costal cartilage to form the superior vena cava. Behind the left brachiocephalic vein and the arch of the aorta, and embraced by the brachiocephalic and left common carotid arteries, is the **trachea.** Over the trachea, and within the pretracheal fascia, the **inferior thyroid veins** pass down to join the left brachiocephalic vein and the thyroidea ima artery from the aorta or brachiocephalic artery may pass upwards to the isthmus of the thyroid. In the young child the left brachiocephalic vein is at a higher level, behind or just above the upper part of the manubrium, and at risk, therefore, during tracheotomy. Immediately behind the manubrium, in front of the great vessels, is part of the **thymus.** The extent and form of the gland varies between individuals and with age. Well developed in the child, it regresses at puberty but may persist into adult life. In front of it descends the thin pretracheal fascia. Arterial branches reach the gland from the internal thoracic arteries. A main, single thymic vein drains to the left brachiocephalic vein. Also between the great vessels and the manubrium may sometimes be found an extension of the **thyroid gland.** An ectopic parathyroid gland may also lie in the superior mediastinum.

113. What lies behind the body of the sternum ?

The posterior surface is covered by a thick membrane formed by the fibres of the radiate ligaments of the sternocostal joints. The **sternocostalis** muscle arises from the margins of the lower third of the bone and the adjacent fifth, sixth and seventh costal cartilages. On each side fibres pass upwards and laterally to the second to the sixth costal cartilages. Sternal branches of the internal thoracic vessels reach into the posterior surface of the bone. The line of junction of the costal **pleura** with the mediastinal pleura on the right passes down near the midline from the manubriosternal joint to the xiphisternal joint. On the left side the line of junction of the two pleural surfaces passes down near the midline from the manubriosternal joint to the level of the fourth costal cartilage, at which level it diverges laterally away from the sternum. Between the manubriosternal joint and the point of divergence of the left pleural reflexion, the two parietal pleural membranes are in contact with each other. Behind the sternum the sharp anterior borders of the **lungs** may follow fairly closely the lines of costomediastinal pleural reflexion, but there is great variability in the size of the costomediastinal pleural recesses. The area of the posterior surface of the body of the sternum not covered by the left pleura is in contact with the fibrous **pericardium**. The **thymus** usually does not extend behind the body of the sternum below the level of the fourth costal cartilages, but may reach the xiphoid process. Well developed up to puberty it subsequently regresses to be replaced by fatty connective tissue, and may be difficult to find in later years. It may retain its form, however, into adult life, as two elongated separate, contiguous, or fused asymmetric lobes, sometimes hard to distinguish from fat, behind the anterior pleural reflexions. The fibrous pericardium lies behind the thymus, and the whole of the posterior surface of the body of the sternum. Within the pericardium the **right atrium** and, below it, the **right ventricle**, lie behind the sternum. The orifices and valves of the heart are behind the sternum. The pulmonary orifice lies horizontally partly behind the left side of the sternum, at the level of the upper border of the third left costal cartilage. The aortic orifice is slightly below and to the right of the pulmonary orifice. Its surface projection passes downwards and to the right behind the left side of the sternum opposite the third intercostal space. The surface projection of the left atrioventricular (mitral) orifice lies obliquely downwards and to the right behind the left side of the sternum opposite the fourth costal cartilage. The surface projection of the right atrioventricular (tricuspid) orifice passes downwards and slightly to the right behind the middle of the sternum, opposite the fourth intercostal space. The valves lie roughly on an oblique line extending downwards and to the right from the level of the third left to the sixth right costal cartilages.

114. How is the manubrium joined to the body of the sternum ?

The manubrium joins the body of the sternum at the manubriosternal joint. The two parts of the sternum meet at an angle of about 15°, forming the prominence that is the angle of Louis, about 5 cm below the jugular notch. In the young adult the articular surfaces are oval and slightly rough. They are joined by a symphysis, like that of the pubis. The bony surfaces are covered by thin layers of hyaline cartilage, which are joined by a layer of fibrocartilage. The structure of the joint varies with age. In the newborn only fibrous and elastic tissue join the parts. Past middle age nearly a third of manubriosternal joints develop a cleft in the fibrocartilaginous disc, so that the joint appears to be synovial. In about 10% of elderly joints ossification develops in the superficial part of the joint, which sometimes proceeds to complete union. There is only very slight movement at the joint with breathing, reportedly not more than about 2° on average in young athletes, though some texts suggest 7°.

115. How are the ribs joined to the sternum ?

Only the upper seven ribs join to the sternum, at its lateral edge, on each side, ending in flattened bars of hyaline cartilage, the costal cartilages. Each first costal cartilage fuses with a superolateral angle of the manubrium at a synchondrosis. The second to the seventh costal cartilages end in fibrocartilaginous surfaces which articulate with similar surfaces on sternal facets at synovial joints. Each second costal cartilage joins the sternum at the adjoining margins of the manubrium and body at the manubriosternal joint. There is often an intra-articular ligament to the manubriosternal cartilage, which then separates the joint into two cavities. Each third, fourth and fifth costal cartilage joins the sternum at a facet opposite a transverse ridge. Each sixth cartilage joins the sternum at a facet in the middle of the lateral border of the lowest segment of the body of the bone. Each seventh cartilage joins the sternum at a facet at the xiphisternal joint. The lower sternocostal joints may have intra-articular ligaments, or they may be synchondroses, especially the seventh. Each sternocostal joint has a thin capsule, reinforced by radiate ligaments, anterior and posterior, the fibres of which spread to mingle with those of joints opposite, above, and below, forming a rather dense membrane covering the bone.

THE RIBS

116. What are the features of a typical rib ?

The third to the ninth ribs are called typical. The first two and the last three have special characteristics.

The features of a typical rib are —

a) the **head,** the slightly expanded and forwardly projecting posterior end, displaying a lower, larger, vertical facet separated from an upper, smaller, upwardly facing facet by a horizontal ridge. The lower facet articulates with the body of the thoracic vertebra of the same number or level, the upper facet with the body of the vertebra above.

b) the **neck,** the flattened part extending laterally from the head to the tubercle, in front of the transverse process of its corresponding vertebra. Its upper border is a sharp ridge, called the **crest** of the neck, for attachment of the superior costotransverse ligament. The lower border is rounder. The anterior surface is smooth and faces anterosuperiorly, the upper part covered by fat separating it from the posterior intercostal membrane, the lower part in direct contact with the pleura. The two areas may be separated by a faint ridge. The back of the neck faces posteroinferiorly, shows vascular foramina and is transversely ridged by the attachment of the inferior costotransverse ligament.

c) the **tubercle,** a prominence on the outer aspect of the junction of the neck and shaft, showing two smooth facets, a medial one covered with articular cartilage for articulation with the costal facet of the transverse process, and a lateral one for attachment of the lateral costotransverse ligament. The articular facets of the upper ribs are convex and face backwards, those of the lower ones are flat and face slightly downwards.

d) the **shaft,** characterised by —

i. bends, in three directions. It is outwardly convexly curved but with a bend near its posterior end, called the **angle**. The posterior end curves upwards and the anterior end downwards. It twists so that posteriorly the inner surface faces slightly upwards, and anteriorly faces downwards.

ii. an outer surface which is smooth. Posteriorly, at the angle, there is a slightly oblique ridge, for attachment of the thoracolumbar fascia over the fibres of iliocostocervicalis. The length of shaft between tubercle and angle increases gradually from above to the eighth rib. A less well defined anterior angle may be present in the lower four typical ribs 3–5 cm from the anterior ends, marked by the sites of interdigitation between the slips of origin of the external oblique, and serratus anterior and latissimus dorsi.

iii. an inner surface which presents the **costal groove** in its posterior two thirds. A prominent ridge continues laterally from the lower border of the neck to form the upper lip of the groove. The lower lip is drawn down to form a prominent flange posteriorly. It fades away anteriorly.

iv. a round upper border.

v. a lower border which is sharp below the costal groove, but round anteriorly.

vi. a cup-shaped anterior end, to receive the costal cartilage.

117. What are the particular features of the first rib ?

The head is small and round, with a single facet for articulation with the body of the first thoracic vertebra only. The neck is narrow, elongated and often directed slightly upwards to the shaft. The tubercle is prominent. There is no angle, the bend of the rib coinciding with the tubercle. The shaft is the shortest and most curved of the ribs. It is flat with upper and lower surfaces, inner and outer borders. There is no costal groove. The anterior half of the upper surface shows oblique grooves for the subclavian vein in front of, and the lower trunk of the brachial plexus and subclavian artery behind, a faint ridge, the inner end of which, at the inner border, forms the prominent scalene tubercle for the insertion of scalenus anterior. Behind the posterior groove is a slightly roughened surface for attachment of scalenus medius, extending behind to the tubercle of the rib. The anterior end is the largest and thickest of the ribs. The first costal cartilage is fixed to the rib by a thin but dense layer of fibrocartilage.

118. What are the relations of the first rib ?

To the **inner border** is attached the suprapleural membrane, or Sibson's fascia. This is a tent-like sheet, sometimes reinforced by muscle, the scalenus pleuralis or minimus, which is attached by its apex to the root of the transverse process of the seventh cervical vertebra, and by its base to the inner border of the rib. It covers the cervical pleura. It is well developed laterally, and is crossed by the subclavian vessels. Medially it fades into the connective tissue of the root of the neck. Anteriorly there is a gap behind the first costal cartilage for the internal thoracic vessels and phrenic nerve. Posteriorly there is a firm border forming the anterior boundary of a gap between it and the neck of the first rib and transverse process of the seventh cervical vertebra. Through this gap pass, from medial to lateral, the stellate ganglion of the sympathetic trunk, the superior intercostal artery, the supreme intercostal vein, and the major component of the ventral ramus of the first thoracic nerve on its way to join the brachial plexus.

To the **outer border**, half way round, is attached part of the first digitation of the serratus anterior muscle, which leaves a small impression. The thin, flat tendon of scalenus posterior crosses the outer border in front of the tubercle, deep to the first digitation of serratus anterior, to reach the second rib.

The **anterior, superior surface** of the first rib is suspended from the cervical vertebral column by scalenus medius and scalenus anterior muscles, and underlies the subclavian vessels and lower trunk of the brachial plexus which cross it. The subclavian artery and lower trunk of the brachial plexus pass through the costoscalene triangle, formed between the scalenus medius, subclavian groove and scalenus anterior. The scalenus anterior is a narrow flat tendon at its insertion. This triangular interval may be the site of compression of the artery and, or, nerve. The resulting symptoms, vascular or nervous, comprise the thoracic inlet (more commonly clinically called outlet) syndrome. The subclavian vein crosses the rib anterior to the attachment of scalenus anterior, and is the frequent site of insertion of an intravenous line. The subclavian vessels become the axillary vessels at the outer border of the rib. The costoclavicular ligament is attached anteriorly to the adjoining rib and costal cartilage, deep to the thick tendon of origin of the subclavius muscle.

The **posterior, inferior surface** of the first rib lies against endothoracic fascia, pleura and lung. The continuation of the first thoracic nerve lies within the outer border.

119. What is the thoracic inlet ?

The thoracic inlet is the upper aperture of the bony thoracic cage, formed by the first thoracic vertebra, the first rib, the first costal cartilage and the upper border of the manubrium. The word inlet is used in anatomical texts, referring perhaps to the entry into the thorax of descending structures, such as the trachea and oesophagus. The same aperture is referred to clinically as the thoracic outlet, referring to the exit from the thorax, to the arm, of the subclavian artery and first thoracic nerve, compression of which at this site gives rise to symptoms. The inlet is suspended by the scalene and sternocleidomastoid muscles. It slopes downwards and forwards at such an angle that the upper border of the manubrium lies opposite the disc between the second and third thoracic vertebrae. It is kidney-shaped, measuring about 5 cm from front to back, and 10 cm from side to side, in an average adult male.

120. What is a cervical rib ?

A cervical rib is an abnormally developed separate costal element which articulates with the seventh cervical vertebra. When such a rib is fully developed there are only six cervical vertebrae, the seventh having the characteristics of a thoracic vertebra. Lesser degrees of development of a cervical rib can occur. The middle part of the rib may be replaced by fibrous tissue. The anterior end of the rib may be attached to the manubrium by ligament instead of costal cartilage. The rib may be shortened, a longer ligament still attached to the manubrium. The short rib may fuse with the rib below, so that the uppermost rib appears bifid. Or it may join the next rib by fibrous tissue only. The junction of the two ribs may form a prominent boss. The costal element of the seventh cervical vertebra may merely be enlarged, and extended, and attached by ligament to the rib below, sometimes as far forwards as the manubrium. The scalene muscles attach to whatever cervical rib presents, or otherwise continue to the rib below.

121. What variations from the normal anatomy of the thoracic inlet may cause symptoms ?

The upper aperture of the thoracic cage is, anatomically, the thoracic inlet. Variations from the normal anatomy of the thoracic inlet may cause angulation and or compression of the first thoracic root or lower trunk of the brachial plexus, and the subclavian artery, causing symptoms of what is known clinically as the thoracic outlet syndrome. Such variations include —
   i. there may be narrowing of the interval between the neck of the first rib and the posterior part of the suprapleural membrane, when ligamentous thickenings of the membrane attach more posteriorly than usual. The lower trunk of the brachial plexus may be affected, particularly the first thoracic nerve fibres.

ii. the first rib may slope so much that the costoscalene triangle is excessively narrowed.

iii. a cervical rib may be present. When a complete cervical rib is present the brachial plexus is prefixed, that is the fourth cervical root makes a major contribution to the upper trunk, and the lower trunk is formed by the seventh and eighth roots. The first thoracic nerve does not then contribute to the plexus. With lesser development of the cervical rib, however, there is increasing contribution from the first thoracic nerve, which must ascend to angulate over it, behind scalenus anterior.

iv. the insertion of scalenus anterior may be unusually high and posterior, or a normally positioned muscle may have a posteriorly extended crescentic attachment ascending under the artery and nerve.

v. the insertion of scalenus medius may extend anteriorly under the artery and nerve and may form a continuous attachment with the insertion of scalenus anterior.

122. How does a rib articulate with the spine ?

From the second to the ninth inclusive, each rib articulates medially by its head with the bodies of its corresponding and upper adjacent thoracic vertebrae, and laterally by its tubercle with the transverse process of its numerically corresponding vertebra.

The features of a **medial, or costovertebral joint** are —

i. the head of the rib which has facets corresponding to the demifacets of the vertebrae, the lower facet the larger and vertical, the upper one smaller and facing upwards. A prominent crest separates the two facets.

ii. the thick intra-articular ligament attached between the crest of the head and the intervertebral disc, separating two synovial cavities.

iii. the thickening of the anterior capsule which forms the triradiate ligament, most apparent in the midthoracic series. From the head fan out upper and lower bands to the corresponding vertebral bodies, and a middle band to the intervening disc, deep to the border of the anterior longitudinal ligament.

iv. the thickening of the posterior capsule which extends medially to meet the laterally directed fibres of the posterior longitudinal ligament on the anulus fibrosus.

The features of a **lateral, or costotransverse joint** are —

i. the small articular facet on the anterior surface of the transverse process.

ii. the small articular facet on the medial part of the tubercle of the rib.

iii. the plane of the articular surfaces, which is concave forward in the upper joints, and rather flatter and facing upwards and forwards in the lower joints, the differences influencing movements of the ribs.

Three **accessory ligaments** support the joints —

i. the **lateral** costotransverse ligament, between the lateral facet of the tubercle and the tip of the transverse process, lateral to the articular facet.

ii. the **inferior** costotransverse ligament, between the back of the neck of the rib and the front of the transverse process, medial to the articular facet.

iii. the **superior** costotransverse ligament, made up of two bands, anterior and posterior, between the crest of the neck and the lower border of the transverse process above. The anterior band passes upwards and laterally. It has a free medial border, but is continuous laterally with the posterior intercostal membrane. The posterior band passes upwards and medially, and blends laterally with the external intercostal muscle.

The head of the **first rib** does not articulate with the seventh cervical vertebra. There is no intra-articular ligament, and so there is only a single synovial cavity. The radiate ligament is less well developed, and there is no superior costotransverse ligament.

The heads of the **twelfth, eleventh,** and often the **tenth ribs** articulate with only one vertebral body, so their joints similarly have single synovial cavities, and less well developed radiate ligaments. The eleventh and twelfth ribs have no costotransverse joints, but do have a few scattered costotransverse fibres. The twelfth rib, near its head, is attached to the transverse process of the first lumbar vertebra by the **lumbocostal ligament** in series with the superior costotransverse ligaments.

123. How does a rib join its costal cartilage ?

The anterior end of a rib is usually depressed to accept the rounded end of the costal cartilage. The periosteum and perichondrium are continuous. There is a thin lamina of fibrocartilage which forms a transition zone between the rib and the hyaline costal cartilage. Sometimes a thin shell of bone extends medially from the rib if the periphery of the costal cartilage ossifies.

124. How do the ribs move ?

There are still uncertainties about the way ribs move at different levels in the thorax, and about the action of the various muscles related to the ribs in breathing. In the upper thorax it can be said that the axis of movement of a rib is a line joining the head of the rib to its tubercle, between the costovertebral and the slightly curved costotransverse joints. This axis is set towards the coronal plane, so that rotation of the rib about it causes the anterior end to rise and fall, without outward displacement. This is the 'pump' action of the rib often described. The costal cartilages undergo slight longitudinal twist, in their substance in the first, and at the sternocostal joints below. They do not alter their direction. The antero-posterior diameter of the upper part of the thorax is increased in inspiration, but the transverse diameter is hardly affected. In the lower thorax the axis of the neck of a rib is more oblique, towards the sagittal plane, so that rotation of a rib about it causes the front of the rib to displace more laterally than forwards. The transverse diameter of the lower part of the thorax is therefore increased in inspiration, with little effect on the anteroposterior measurement. At the same time, because the costotransverse joints are flatter in the lower than in the upper thorax, sliding movements occur at their surfaces, so that a 'bucket handle' movement is added to the pump action described for the upper ribs. The costal margin cartilages become more horizontal, with slight translation movement at the interchondral joints and widening of the infrasternal angle.

125. What muscles are attached to the outer aspect of the thorax ?

The muscles that are attached to the outer aspect of the thorax can be listed in three groups of four —

Anteriorly —

Subclavius, to the upper surface of the anterior end of the first rib and adjacent first costal cartilage.

The **sternocostal part of pectoralis major,** to the anterior end of the sixth rib, the second to the sixth costal cartilages, the medial half of the adjacent sternum and manubrium and part of the aponeurosis of the external oblique.

Pectoralis minor, to the upper margins and outer surfaces of the third, fourth and fifth ribs near their cartilages.

Rectus abdominis, by its most lateral fibres to the anterior end of the fifth rib, and by its remaining fibres to the costal cartilages of the fifth, sixth and seventh ribs and the side of the xiphoid process.

Laterally —

Scalenus posterior, to the outer border of the second rib, behind the tubercle for serratus anterior.

Serratus anterior, to the outer surfaces and upper borders of the upper eight ribs. The first digitation arises from the first and second ribs, and the intervening intercostal fascia. The remaining digitations arise from the numerically corresponding ribs in an oblique line passing downwards and backwards. The lower four digitations fit between those of the origin of the external oblique.

External oblique, to the outer surfaces and lower borders of the lower eight ribs, in an oblique line passing downwards and backwards, interdigitating with the slips of origin of the serratus anterior above and latissimus dorsi below. The upper digitations are attached close to the costal cartilages, the middle ones some way behind, and the lowest to the tip of the twelfth costal cartilage.

Latissimus dorsi, to the lower three or four ribs, interdigitating with the lower slips of origin of the external oblique.

Posteriorly —

Serratus posterior superior, to the upper borders and outer surfaces of the second, third, fourth and fifth ribs, a short distance lateral to their angles.

Serratus posterior inferior, to the lower borders and outer surfaces of the lower four ribs, a short distance lateral to their angles.

Levatores costarum, to the upper borders and outer surfaces of the ribs, between their tubercles and angles. They arise from the ends of the transverse processes of C7 to T11 vertebrae. The first muscle is fused with scalenus medius. They pass parallel to the posterior borders of the external intercostals.

Parts of erector spinae, to the angles of all the ribs (iliocostocervicalis) and to between the tubercles and angles of the lower ribs (longissimus thoracis).

126. How are the intercostal muscles arranged ?

The intercostal muscles form three thin, but extensive layers between adjacent ribs.

  i. An **external intercostal** muscle extends from the sharp inferior border of a rib to the round superior border of the rib below. Posteriorly it blends with the posterior fibres of the superior costotransverse ligament. Anteriorly, between the costal cartilages, it forms the aponeurotic external intercostal membrane. Its fibres pass diagonally downwards from an attachment nearer the spine to an attachment nearer the sternum, so that, on geometric grounds, contraction of its fibres raises the ribs, i.e. it is an inspiratory muscle.

  ii. An **internal intercostal** muscle extends from the costal groove of a rib, and its costal cartilage, to the superior border of the rib and costal cartilage below. Anteriorly it reaches the sternum. Posteriorly it forms the aponeurotic internal intercostal membrane, between the angles of the ribs laterally and the anterior fibres of the superior costotransverse ligament medially. Its fibres pass diagonally downwards in an opposite direction to the external intercostal muscles, from an attachment nearer the sternum to an attachment nearer the spine. On geometric grounds contraction of its fibres between the costal cartilages raises the ribs, i.e., they are inspiratory, and of those between the ribs lowers them, i.e., they are expiratory.

  iii. An **innermost intercostal** muscle is present only in the lower part of the chest wall. It extends between the inner surfaces of the middle two quarters of adjacent ribs. The fibres run in the same direction as the internal intercostal muscle, and they blend posteriorly with a subcostalis. On geometric grounds they would assist in depression of the ribs.

However, the action of the intercostal muscles appears to be complex and not fully understood, their activity seeming to vary in different parts of the chest wall and in different types of respiratory effort.

127. What muscles are attached to the inner aspect of the thorax ?

The muscles that are attached to the inner aspect of the thorax are —
    The innermost intercostals (see question 126).

Transversus abdominis, in part to the inner aspects of the lower six costal cartilages, interdigitating with the slips of origin of the diaphragm.

Transversus thoracis (sternocostalis). It ascends, as a thin layer, from the back of the lower third of the body of the sternum and xiphoid process to the lower borders and inner surfaces of the second to the sixth costal cartilages. The lower fibres pass horizontally in contact with those of transversus abdominis. The middle fibres run obliquely and the highest are vertical. The middle and higher fibres may possibly help raise the ribs, by reducing the inferior angles between the costal cartilages and the sternum. The term transversus thoracis is sometimes used to include the innermost intercostal and subcostal muscles.

Subcostales. They lie in the posterolateral part of the chest, best developed over the lower ribs, parallel with the fibres of the internal intercostals. The lowest arises from the inner surface and upper border of the last rib, and inserts onto the inner surface of the tenth rib. The higher fibres also bridge over one or two ribs. The highest extends usually from the fourth to the second rib. They form a thin sheet in the paravertebral gutter.

Sternohyoid and **sternothyroid** (see question 110).

The **diaphragm** (see question 128).

128. What are the skeletal attachments of the diaphragm ?

The periphery of the hemidome of the diaphragm is attached from front to back to —

　i. the back of the xiphoid process and xiphisternal joint.

　ii. the inner aspect of the costal cartilages and adjacent bony ends of the seventh to the twelfth ribs.

　iii. the **lateral arcuate ligament**, the thickening of the anterior lamina of the thoracolumbar fascia which stretches across quadratus lumborum from the twelfth rib to the tip of the transverse process of the first lumbar vertebra.

　iv. the **medial arcuate ligament**, the tendinous arch over psoas major, which stretches from the tip of the transverse process of the first lumbar vertebra to merge with the lateral tendinous border of a crus of the diaphragm.

　v. the bodies and intervening intervertebral discs of the upper three lumbar vertebrae on the right and two on the left, forming the **crura** of the diaphragm. The tendinous fibres merge medially with the anterior longitudinal ligament, and may meet in the midline.

129. What are the muscles of inspiration ?

Primary inspiratory muscles are those used in gentle breathing. They include the diaphragm pre-eminently, assisted by the external intercostals, anterior internal intercostals, transversus thoracis and levatores costarum. **Accessory** inspiratory muscles are those used in forced breathing. They are the sternocleidomastoids, scalenes, costocervicales, posterior superior serrati, and, acting from a fixed, abducted arm, the pectorals.

130. What are the muscles of expiration ?

Primary expiratory muscles are those used in gentle breathing, to enhance the natural elastic expiratory recoil of the chest wall. They comprise the posterior fibres of the internal intercostal muscles. **Accessory** expiratory muscles are those used in forced breathing, and in singing, shouting and coughing. They are the abdominal muscles (rectus abdominis, external and internal obliques and transversus), quadratus lumborum, serratus posterior inferior, iliocostalis and longissimus .

131. What is attached to the twelfth rib ?

To the **inner surface** are attached:

The lowest **subcostalis** muscle, to the upper part of nearly the whole length of the rib, to pass upwards and laterally in the same direction as the internal intercostal fibres.

Quadratus lumborum, to the lower part of the medial half to two thirds of the surface, between the attachment of the lumbocostal ligament medially to the origin of transversus laterally.

The anterior layer of the **thoracolumbar fascia**, which defines the upper limit of attachment of quadratus lumborum. The lateral arcuate ligament is a thickening of the fascia which passes across the muscle from the tip of the first lumbar transverse process to the lateral limit of attachment of the muscle to the rib, where the aponeurotic origin of transversus abdominis begins.

The **diaphragm**, to the upper part of the surface lateral to the attachment of the lateral arcuate ligament.

The **transversus abdominis** to the lower part of the lateral end.

To the **outer surface** are attached:

Tendinous fasciculi of **longissimus thoracis, iliocostalis lumborum** and **iliocostalis thoracis** to the medial end.

The posterior lamina of the **thoracolumbar fascia**, covering erector spinae, to the lateral margin of the medial area.

The lowest fasciculus of the **serratus posterior inferior** to the lower part of the surface just lateral to the attachment of the thoracolumbar fascia.

The lowest costal digitation of **latissimus dorsi** from the upper part of the tip.

The lowest costal digitation of the **external oblique** from the lower part of the tip.

To the **upper border** are attached:

Scattered fibres of rudimentary costotransverse ligaments to the most medial part.

The lowest intercostals (external, and internal and posterior intercostal membrane).

Tendinous fasciculi of the lowest **levator costae**, from the ligamentous area to the attachment of the thoracolumbar fascia.

To the **lower border** are attached:

The strong **lumbocostal ligament** medially, from the tip of the transverse process of the first lumbar vertebra, in line with the thoracic superior costotransverse ligaments.

The middle lamina of the **thoracolumbar fascia**, behind quadratus lumborum, lateral to the lumbocostal ligament.

The aponeurotic origin of **transversus** and **internal oblique** lateral to **quadratus lumborum**, with muscle fibres of internal oblique near the tip.

132. What is the relationship of the twelfth rib to the pleura ?

The costodiaphragmatic recess of pleura spreads across the medial part of the anterior surface of the rib. The line of reflexion crosses the anterior surface of the rib at about a point opposite to the attachment of the posterior lamina of the thoracolumbar fascia to the posterior aspect of the rib. It passes medially and horizontally on a level midway between the rib and the transverse process of the first lumbar vertebra to below the head of the rib, where it becomes continuous with the mediastinal pleura. The anterior surface of the rib between the point of crossover of the line of reflexion and the head, is covered partly by subcostalis above and quadratus lumborum below. The recess of pleura is separated from the kidney and its surrounding fat and fascia anteriorly, by the origin of the diaphragm from the lateral arcuate ligament.

133. How are the intercostal nerves and vessels related to the ribs ?

Forward of the angles of the ribs, in each **typical intercostal space**, there is a main neurovascular bundle which passes round the chest wall between the internal intercostal muscle and the innermost plane of muscle formed by the subcostales and intercostales intimi. The nerve is a thoracic anterior primary ramus, the intercostal nerve. The vessels are the communicating anterior and posterior intercostals. They lie deep to the costal groove where present, the vein above, the nerve below and the artery in between. The lateral and anterior cutaneous branches pass out between the ribs. The lateral branches form a line of exit from the spaces, curving gently downwards and forwards from the posterior axillary border above, to the anterior border below. Just medial to the angles collateral branches are given off and journey in the same plane as the main bundle, just above the upper border of the rib below. Medial to the angles of the ribs, the neurovascular bundles lie between the ribs, anterior to the superior costotransverse ligaments and anterior intercostal membranes.

In the first space the major part of the first thoracic nerve crosses the neck of the first rib, lateral to the superior intercostal artery, to join the eighth cervical nerve to form the lower trunk of the brachial plexus. The minor continuation of the ramus courses round within the outer border of the rib, in contact with the endothoracic fascia and pleura. The first intercostal artery is a branch of the superior intercostal artery which arises from the costocervical trunk and crosses the neck of the rib. The first, or supreme intercostal vein also crosses the neck of the rib medial to the artery, to join the brachiocephalic or vertebral vein.

In the second space the posterior intercostal artery arises from the superior intercostal artery, which crosses the neck of the second rib.

On the right **the posterior intercostal veins** of the second, third, and fourth spaces cross the ribs to form the right superior intercostal vein, which joins the azygos vein. On the left the posterior intercostal veins of the second, third, and sometimes the fourth spaces also cross ribs to form a superior intercostal vein which then leaves the thoracic wall to cross the left side of the arch of the aorta, lateral to the vagus nerve, medial to the phrenic nerve, to join the left brachiocephalic vein.

The seventh and eighth intercostal nerves on each side pass medially **behind the costal margin**, between the digitations of the diaphragm and transversus abdominis, to reach between the transversus and the posterior layer of the aponeurosis of the internal oblique, medial to the lateral border of rectus abdominis, to enter the sheath from behind. The ninth nerve likewise passes behind the costal margin, and pierces the lateral margin of the rectus sheath. The tenth and eleventh nerves, above and below the floating eleventh rib pass in the same muscular planes, but do not have a costal cartilage to cross behind. The subcostal nerve and vessels pass along the lower border of the twelfth rib, anterior to quadratus lumborum, under the lateral arcuate ligament, before piercing the origin of transversus.

The upper six pairs of anterior intercostal vessels arise from the internal thoracic vessels which descend **behind the costal cartilages**, a finger's breadth lateral to the sternal margin. The lower three arise from the musculophrenic vessels which slope down behind the seventh and eighth costal cartilages, to pierce the diaphragm near the ninth cartilage.

134. What ribs relate to the costodiaphragmatic reflexion of pleura ?

The line of costodiaphragmatic pleural reflexion on the right begins anteriorly behind the xiphisternal joint and descends laterally to behind the eighth rib in the midclavicular line, and the tenth rib in the midaxillary line. It then passes medially and slightly upwards across the twelfth rib at the outer border of the erector spinae, to level with the upper border of the spine of the twelfth thoracic vertebra about 2.5 cm from the midline where it meets the mediastinal pleura. The line of reflexion on the left side begins at a variable point lateral to the left border of the sternum (up to about 2.5 cm), at the level of the xiphisternal joint, and passes around the chest wall in a similar way to that on the right side, though slightly lower.

135. Which ribs relate to the inferior border of the lung ?

The inferior border of each lung is about 5 cm above that of its costodia-phragmatic reflexion, leaving a costodiaphragmatic recess of pleura, filled only on full inspiration. The lower border of the quietly inspiring lung crosses the eighth rib in the midaxillary line, and the tenth rib posteriorly, to level with the tenth thoracic spine 2.5 cm from the midline. In the midaxillary line it is about 10 cm above the costal margin.

136. What ribs relate to the oblique fissures ?

On the left side the oblique fissure can be drawn as a line extending from a point level with the interspace between the spines of the third and fourth thoracic vertebrae, roughly along the vertebral border of the abducted scapula, across the fifth rib in the midaxillary line to the sixth costochondral junction. On the right side the oblique fissure is slightly more horizontal, beginning posteriorly opposite or just below the fourth thoracic spine, and ending behind the sixth costochondral junction.

137. What rib relates to the transverse fissure of the right lung ?

The transverse fissure of the right lung lies behind the anterior part of the fourth rib. Anteriorly it extends to behind the fourth costal cartilage and the lateral part of the sternum, where it meets the anterior margin of the lung. Laterally it extends to the midaxillary line where it meets the oblique fissure.

138. Where is it safest to insert a pleural drainage tube into the thorax ?

When inserting a chest drain it should be borne in mind that —

i. the pleural cavity does not extend lower than the tenth rib in the midaxillary line.

ii. the internal thoracic vessels pass down behind the costal cartilages about a finger's breadth from the sternal margin.

iii. posteriorly, medial to the angles of the ribs, the intercostal vessels lie in the middle of the spaces.

iv. forward of the angles of the ribs the vessels lie just above the lower border of the rib above, but a collateral vessel lies just above the upper border of the rib below.

v. the long thoracic nerve runs down under the fascia over serratus anterior, to supply its digitations, behind the midaxillary line.

A drain should not be put in within 2.5 cm of the sternal margin, or medial to the posterior angles of the ribs or the outer border of erector spinae, for fear of injuring the internal thoracic or posterior intercostal vessels. Laceration of these vessels may result in severe bleeding into the pleural cavity. A drainage hole should be made midway between two ribs, or towards the lower part of a space, to avoid damage to the main vessels deep to the upper rib. Thoracic surgeons describe a **triangle of safety** for intubation of the chest wall, to avoid damage to thoracic or abdominal contents. The area is outlined by the anterior and posterior axillary borders, and the sixth rib, level with the nipple in men, and the lower border of the base of the breast in women. The midaxillary line, in the area, is the favoured site, the point entering the space midway between the ribs.

139. Where is it safest to aspirate a pericardial effusion ?

A needle passed through the precordium lateral to the lower lateral margin of the sternum may be dangerous for two reasons. The line of pleural reflexion behind the sternum below the fourth costal cartilage on the left side may not diverge laterally beyond the lateral border of the sternum until it reaches the level of the xiphisternum, so that a needle passed through the sixth space beside the lateral border of the sternum could puncture pleura and lung on its way to the pericardial cavity. Also the costal cartilages of the sixth and seventh spaces may be fused together towards their insertion, so that an entering needle between them is forced laterally, and possibly through the pericardiophrenic vessels. It is best to enter a needle into the pericardium upwards just lateral to the xiphoid process.

THE UPPER LIMB

THE CLAVICLE

140. How is the clavicle orientated ?

The medial two thirds of the clavicle are convex forwards, roughly cylindrical, and end medially in a quadrangular, slightly saddle-shaped articular surface. The lateral third is concave forwards, flattened with superior and inferior surfaces, and ends laterally in a small oval articular surface. The undersurface of the bone displays, from medial to lateral ends, the rhomboid impression for the costoclavicular ligament, the impression for the subclavius muscle, the conoid tubercle and the trapezoid ridge.

141. What does the clavicle do ?

  i. The clavicle provides a strut to absorb the thrust of the arm through, and resist traction of the arm from, the shoulder.

  ii. It gives attachment to pectoralis major, deltoid, trapezius, and subclavius, muscles that manoeuvre the shoulder complex.

  iii. It protects underlying important structures such as the brachial plexus, subclavian and axillary vessels, first rib and apex of the lung, from direct trauma.

iv. It holds the shoulder backwards, upwards and outwards in its natural position. If the clavicle is congenitally absent or is removed surgically because of chronic infection, tumour or troublesome fractures the shoulder tends to droop inwards and downwards, and is said to tire more easily.

142. How does the clavicle develop ?

Like all long bones the rudiment of the clavicle is a band of condensed mesenchyme. In all long bones except the clavicle the mesenchyme progresses to a cartilaginous model before ossification begins. The clavicle is unique in that it seems that ossification commences before the cartilaginous model has time to develop. Between the fifth and sixth weeks of fetal life medial and lateral centres of intramembranous ossification develop in the middle of the mesenchymal band. They fuse together within a week. By the time the cartilaginous model develops at either end ossification in the middle of the bone is well advanced. Ossification proceeds from the centre into the cartilaginous zones. Fronts of endochondral ossification develop and growth in length continues as in any long bone. A secondary centre for the medial end appears in the late teens. It forms a thin disc which is not easy to see on routine X-ray films, and when it is visible may be rather irregular in appearance and can mistakenly be thought pathological. It does not fuse with the shaft until nearly twenty five years of age. It is important to recognise the existence of this centre of ossification because an apparent dislocation of a sternoclavicular joint may in fact be a fracture-separation of the medial epiphysis. The medial end of the clavicle contributes most of the growth in length of the bone. An ossification centre for the lateral end also occurs at about the same time as the medial end, but is seen only rarely on X-ray films. The primary centre of ossification of the clavicle is the first to appear in the body, while the secondary centre for the medial end is possibly the last to appear and the last to fuse.

143. How does the clavicle articulate with the sternum ?

The sternal end of the clavicle articulates with the clavicular notch of the manubrium, and sometimes with a small adjoining part of the first costal cartilage. There is a firm capsule, stronger in front than behind. The sternoclavicular joint is not primarily designed to bear load, as can be deduced from several of its interesting features. Thus less than half of the articular surface of the clavicle covers that of the manubrium, making it potentially the least stable joint in the body. Although both surfaces are slightly saddle-shaped they are not congruent. The articular surfaces are covered by fibrocartilage rather than hyaline cartilage. The joint is completely divided by a fibrous disc attached strongly to the junction of the first costal cartilage and the manubrium below, and to the upper and posterior margins of the clavicular articular surface above, as well as to the capsule in front and behind.

144. How does the clavicle articulate with the scapula ?

The clavicle articulates with the acromion at the acromioclavicular joint. The articular surfaces are relatively small, oval areas, covered by fibrocartilage. There is variation in the plane of the joint. There is often an override of the clavicle on the acromion of up to an angle of about 50°, but the plane of the joint may be vertical, or the acromion may override the clavicle. There is an articular disc which may be complete or meniscoid. It degenerates by middle age. The capsule is thin but is reinforced above by the attachments of deltoid and trapezius. There is only a very limited range of movement in the joint. There is an incidence of about 1% of actual articulation between the coracoid process and the conoid tubercle.

145. What movements can the clavicle make ?

At the sternoclavicular joint movement of the clavicle can occur in all directions, the medial end of the bone hinging on the costoclavicular ligament. Movements of the shoulder therefore occur in opposite directions to the medial end of the clavicle. The outer clavicle can elevate about 35°, and protract (forwards) and retract (backwards) through a total angle of about 35°. The clavicle can rotate on its long axis, like a cam shaft, about 45°. Rotation of the clavicle is an indispensable adjunct to full abduction of the arm.

At the acromioclavicular joint very little movement occurs. It has been shown, using adjacent wire pointers drilled into the clavicle and the acromion of volunteers that only between 5° and 8° of deviation of the pins occur in any direction. The clavicle and scapula move in an almost synchronous manner.

146. What strains are applied to the sternoclavicular joint ?

With movements of the shoulder the medial end of the clavicle can be displaced upwards, forwards, backwards, downwards, medially or laterally, or be rotated to face upwards or downwards on its longitudinal axis. All these movements of the clavicle hinge on the costoclavicular ligament.

Upward displacement of the medial end has been shown by experiment on cadavers to be resisted primarily by the capsule of the joint. The articular disc plays a relatively minor part in resistance to this movement.

Anterior and **posterior displacement** of the medial end is also resisted by the capsule, anterior displacement more effectively than posterior.

Downward displacement of the medial end is resisted by impaction against the first costal cartilage.

Lateral displacement of the medial end is resisted by the anterior fibres of the costoclavicular ligament, which pass upwards and laterally from the anterior part of the medial end of the first rib to the anterior part of the rhomboid impression of the clavicle. They are separated from the posterior fibres of the ligament by a small bursa. The interclavicular ligament also tightens when the arm carries weight.

Medial displacement of the medial end is resisted by the articular disc and the posterior fibres of the costoclavicular ligament which pass upwards and medially from the first rib a little lateral to the anterior part of the ligament to the posterior part of the rhomboid impression of the clavicle.

Upward rotation of the clavicle is resisted by the anterior fibres of the costoclavicular ligament.

Downward rotation is resisted by the posterior fibres of the costoclavicular ligament.

147. On what does the security of the acromioclavicular joint depend ?

The chief stabilising factor for the acromioclavicular joint is the **coracoclavicular ligament**. It has two components—

    i. the **conoid** part, a strong triangular band attached by its apex to the posteromedial side of the root of the coracoid process. It sweeps up almost vertically to insert by its base into the conoid tubercle of the clavicle at the point of maximum anterior concavity of the lateral part of the bone.

    ii. the **trapezoid** part, a thin quadrilateral band arising from the upper surface of the horizontal part of the coracoid process anterolateral to the origin of the conoid ligament. It passes almost horizontally to insert into the trapezoid ridge of the clavicle, which extends anterolaterally from the conoid tubercle.

The coracoid process of the scapula hangs just about 1 cm below the clavicle. The coracoclavicular ligament provides a very strong restraint against the acromion collapsing downwards and medially below the outer end of the clavicle, as happens when the ligament ruptures after impact trauma.

148. What muscles are attached to the clavicle ?

The clavicular head of **sternocleidomastoid** arises from the superior aspect of the medial end.

The clavicular fibres of **pectoralis major** arise from the anterior aspect of the medial two thirds.

The clavicular fibres of **deltoid** arise from the anterior aspect of the lateral third.

The clavicular fibres of **trapezius** insert into the posterior aspect of the lateral third.

Subclavius inserts into the inferior surface of the middle third, from the impression for the costoclavicular ligament to in front of the conoid tubercle.

Part of **sternohyoid** arises from the posterior aspect of the capsule of the sternoclavicular joint and the medial end of the clavicle.

149. What lies close behind the clavicle ?

i. Behind the medial end, at the sternoclavicular joint, is the origin of sternohyoid and passing up behind this is the lower part of sternothyroid. Behind these muscles the subclavian and internal jugular veins unite to form the brachiocephalic vein and to receive the terminations of the lymphatic trunks on the right and the thoracic duct on the left. The anterior jugular vein passes laterally between sternocleidomastoid and sternohyoid just above the joint to empty into the external jugular or brachiocephalic veins. It may be large and valveless, bleeding profusely if damaged at surgery. Behind the confluence of veins on the right the brachiocephalic artery divides into the right subclavian and common carotid arteries. On the left the common carotid and subclavian arteries diverge. The vagus nerve passes in front of the first part of the right subclavian artery giving off its recurrant laryngeal branch to ascend behind the artery. The left vagus nerve is posterolateral to the left common carotid artery.

ii. **Behind the forwardly curved medial two thirds**, the clavicle, together with the coracoid process and the upper border of the scapula, and the first rib, form the margins of the **cervico-axillary** canal. Medially, passing down through this canal, is the subclavian vein. It passes in front of the insertion of the scalenus anterior into the first rib, the phrenic nerve intervening. Lateral to the vein the third part of the subclavian artery emerges from behind the insertion of scalenus anterior. Lateral to the artery the trunks of the brachial plexus, coming from the posterior triangle of the neck, bifurcate into their divisions to form the cords in the upper axilla. The subclavian artery and the trunks of the plexus are enclosed by a continuation of prevertebral fascia to form the axillary sheath. The nerve to subclavius and the suprascapular vessels lie behind the bone. The dome of the cervical pleura, covered by the suprapleural membrane (Sibson's fascia) rises up 2.5 cm above the medial third of the clavicle, behind the subclavian vessels and scalenus anterior.

iii. **Behind the outer third** of the clavicle the supraspinatus muscle emerges from the supraspinatus fossa.

THE SCAPULA

150. What is the anatomical position of the scapula ?

The medial or vertebral border is nearly vertical, though the upper end is a little nearer the midline than the lower. The upper angle is level with the spine of the second thoracic vertebra. The triangular area at the medial end of the crest of the spine of the bone is level with the spine of the third thoracic vertebra. The inferior angle is level with the spine of the seventh thoracic vertebra. The lateral or axillary border is at an angle of about 50° to the horizontal. The body of the bone, encased in muscle, is held against the chest wall at an angle of between 30–40° to the coronal plane, with its costal surface facing very slightly downwards in conformation with the curvature of the thoracic cage.

151. What parts of the scapula can be felt in the living ?

The tip of the coracoid process is palpable below the outer third of the clavicle, just lateral to the infraclavicular fossa. The whole length of the crest of the spine of the scapula and the acromion and its angle are subcutaneous. The inferior angle can nearly always be felt, even in obese subjects, especially if the shoulder is moved. The medial and lateral borders can be felt less distinctly through muscle.

152. Identify the coracoid process. What is attached to it ?

The coracoid process is the hook-like projection from the summit of the head of the scapula. Its short, thick root projects upwards and slightly forwards to the angle of the process. The horizontal part curves forwards, laterally and a little downwards from the angle.

Attached to the coracoid process are —

a) **three** tendons —

i. the tendon of **pectoralis minor**, to the medial border and upper surface of the horizontal part. Part of the tendon may pass into the coracoacromial ligament, or through it to become continuous with the coracohumeral ligament.

ii. the combined tendons of **coracobrachialis**, medially, and the short head of **biceps**, laterally, to the tip of the horizontal part.

b) **three** ligaments —

i. the **coracoclavicular ligament**, consisting of two parts. The conoid ligament is attached to the posteromedial side of the root. The trapezoid ligament is attached to a line running anterolaterally from the angle along the upper surface of the horizontal part of the process.

ii. the **coracoacromial ligament** is attached to the length of the lateral aspect of the horizontal part of the process.

iii. the **coracohumeral ligament** arises from the lateral aspect of the root of the process just above the supraglenoid tubercle.

153. Demonstrate how the clavicle normally lies in relation to the scapula.

With the two articular surfaces of the acromioclavicular joint apposed, the conoid tubercle of the clavicle is about a centimetre above the root of the coracoid process. These are the sites of attachment of the conoid part of the coracoclavicular ligament.

154. Identify the suprascapular notch. What passes through it ?

The suprascapular notch indents the lateral part of the upper border of the bone, bounded laterally by the sharp medial border of the root of the coracoid process. It defines the upper boundary of the surgical neck of the scapula, the anatomical neck being close to the rim of the glenoid cavity. It is converted by the upper transverse scapular ligament into a foramen through which passes the susprascapular nerve on its way to supply the supraspinatus and infraspinatus muscles. The suprascapular vessels pass above the ligament.

155. What movements can the scapula make ?

The scapula, surrounded by muscles, is mobile on the chest wall, at what can be called the **scapulothoracic joint,** an important part of the shoulder complex of joints. The scapula may undergo —

   i. elevation and depression, without rotation.

   ii. protraction and retraction, that is forward and backward displacement, away from and towards the spine, around the chest wall.

   iii. rotation, so that the glenoid surface faces obliquely upwards or downwards.

156. What muscles are prime movers of the scapula and where are they attached ?

Elevation results from contraction of —

   i. the upper fibres of trapezius, inserted into the medial margin of the acromion and the superior lip of the crest of the spine, recurving a little around the tubercle at the apex of the triangular area at the medial end of the crest of the spine.

   ii. levator scapulae, inserted into the medial border between the superior angle and the upper border of the triangular area.

Depression results from contraction of —

   i. the lower fibres of trapezius.

   ii. some fibres of latissimus dorsi inserted into the posterior aspect of the inferior angle, as most of the fibres of the upper margin of the muscle sweep over the bone.

Protraction results from contraction of —

   i. serratus anterior, inserted into the costal surfaces of the superior angle, medial border and inferior angle.

   ii. pectoralis minor, inserted into the medial border and superior surface of the horizontal part of the coracoid process.

Retraction results from contraction of —

   i. trapezius.

   ii. rhomboids, minor and major. Rhomboid minor is attached to the medial border of the triangular area at the medial end of the crest of the spine. Rhomboid major is attached to the medial border between the lower limit of the triangular area and the inferior angle. This latter attachment is by the upper and lower tendinous bands, with a thin membrane intervening.

Upward rotation results from contraction of —

   i. the upper and lower fibres of trapezius.

   ii. the lower fibres of serratus anterior.

Downward rotation results from contraction of —
   i. levator scapulae.
  ii. rhomboid minor and major.
 iii. pectoralis minor.

157. What is a 'winged scapula' ?

The word denotes an abnormal prominence of the inferior angle and medial border of the scapula. The most obvious prominence follows paralysis of serratus anterior (long thoracic nerve, C5, 6 and 7). Slight abnormal prominence of the inferior angle will follow paralysis of the trapezius (accessory nerve) or paralysis of the rhomboids (dorsal scapular nerve, C5).

158. What muscles move the arm from the scapula and where do they arise from the bone ?

Supraspinatus arises from the medial two thirds of the supraspinatus fossa. It assists deltoid to abduct the humerus.

Infraspinatus arises from the medial two thirds of the infraspinatus fossa. It laterally rotates the humerus.

Teres minor arises from an elongate area along the upper two thirds of the lateral side of the dorsal surface. It laterally rotates the humerus. It is also a weak adductor.

Teres major arises from an oval area over the dorsal surface of the inferior angle. It medially rotates and adducts the humerus.

Subscapularis arises from the medial two thirds of the subscapular fossa. It medially rotates the humerus.

The above muscles all assist each other in steadying the upper end of the humerus against the glenoid during movements of the arm. They are enclosed in fasciae attached to the margins of the bone; that over the infraspinatus and teres minor is particularly strong. Muscle fibres arise from the deep aspects of the fasciae. When the scapula is fractured, bleeding occurs into tight compartments and causes considerable pain. The confined haematoma results in a characteristic swelling. At the same time the musculofascial envelope serves to splint the fracture fragments, and so prevents their displacement.

Deltoid arises from the inferior lip of the crest of the spine, and the lateral border of the acromion. Anterior fibres arise from the anterior surface of the lateral third of the clavicle. Working alone, deltoid pulls the humerus upwards against the coracoacromial arch. In combination with the rotator cuff muscles it abducts the humerus. The posterior fibres take part in extension and lateral rotation and the anterior fibres in flexion and medial rotation.

The long head of **biceps** arises from the supraglenoid tubercle, within the capsule of the shoulder joint. Mainly a flexor of the elbow and supinator of the forearm, the biceps is also a flexor of the humerus. With the humerus laterally rotated and the elbow extended, the biceps can assist in abduction of the arm.

The long head of **triceps** arises from the infraglenoid tubercle, outside the capsule of the shoulder joint. It extends the elbow.

159. How does the scapula develop ?

At birth there is only one centre of ossification, for the body of the bone. The coracoid process develops from a centre which appears in the middle of it during the 1st year, and then by another centre, called the subcoracoid centre, which forms in the base of the coracoid process and upper quarter of the glenoid fossa at about puberty or shortly afterwards. There may be small centres of ossification at the tip or in the dorsal part of the root of the coracoid process. Other centres which appear at puberty are up to five for the acromion, one along the medial border and inferior angle, and a horseshoe shaped one for the lower three quarters of the glenoid. All these centres fuse in the early 20s. The centres for the coracoid process and · acromion are sometimes mistaken for fractures.

160. Where is the capsule of the shoulder joint attached ?

It arises from the rim of the glenoid cavity except superiorly where the capsule encloses the origin of the long head of biceps from the supraglenoid tubercle. The infraglenoid tubercle, for the long head of triceps, lies outside the capsule.

161. What are the features of the glenoid cavity ?

The glenoid cavity, known clinically simply as the glenoid, is the articular surface of the scapula, for the head of the humerus, part of the glenohumeral joint. It is a shallow rather pear-shaped cavity. A slight indentation of its upper anterior border gives it an appearance suggestive of an inverted comma. The cavity is deepest in the centre of the lower part, or body of the comma, where the articular cartilage is at its thinnest. Relative to the axis of the scapula, which is taken to be the line of junction of the spine with the blade, the surface of the glenoid tilts slightly downwards about 5°. The surfaces of most glenoids also face slightly backwards, at an average of about 7°. Some glenoid surfaces face slightly forwards, however, and it has been reported that such anterior tilt of the glenoid is a common finding in cases of recurrent anterior dislocation, and it is therefore suggested as a possible factor predisposing to anterior instability of the joint.

162. What is the labrum of the glenoid ?

The labrum is a rim of dense fibrous tissue attached to the margin of the glenoid at the meeting point of articular cartilage and periosteum. In radial section it is a triangular structure with its base attached to bone by a narrow transitional zone of fibrocartilage. It is up to 0.5 cm wide above and below, usually slightly less in front and behind. The glenoid surface of the labrum is continuous with the articular surface of the glenoid, extending the socket for the head of the humerus, but adding little to its stability. Its outer surface and periphery are continuous with the capsule. The relaxed capsule allows the peripheral rim of the labrum around the lower half of the glenoid, corresponding to the body of its inverted comma-like surface, to pout slightly so that synovium covers the outer surface of its edge, forming a shallow recess behind it. With rotation of the head and tightening of the capsule, the rim loses its prominence and the recess disappears. Thus lateral rotation eliminates the anterior labral rim and its recess, and medial rotation eliminates the posterior rim and its recess. The upper part of the labrum gives attachment to the long head of the biceps. As well as arising from the supraglenoid tubercle the tendon of tne long head blends with the labrum, curving into its posterosuperior quadrant. It may be that traction by the labral origin of biceps is a causative factor in the detachment of this part of the labrum commonly seen in later decades. Detachment of the labrum and capsule from the anterior rim of the glenoid is an important causative factor in recurrent anterior dislocation of the shoulder joint.

THE HUMERUS

163. How does the alignment of the head and anatomical neck of the humerus compare with that of the femur ?

Like the femur, the axis of the head and neck of the humerus is at an angle of inclination of about 125° to the longitudinal axis of the shaft. **In contrast** to the femur the axis of the head and neck of the humerus is at an angle of **retroversion** of about 30–40° to the interepicondylar line. In the anatomical position this aligns the head and neck of the humerus with the axis of the scapula (the line of the junction of the spine and the blade). The angle of retroversion can vary to some degree between individuals and in the same individual on the two sides.

164. Where is the capsule of the shoulder (glenohumeral) joint attached to the humerus ?

The true capsule of the shoulder joint is a relatively thin structure which attaches to the anatomical neck of the bone, that is to immediately beyond the articular surface, over its upper, anterior and posterior aspects. Below, where it is lax, it descends to the surgical neck 1–2 cm below the articular margin. This part of the capsule is known as the axillary pouch.

165. What strengthening ligaments are there of the capsule of the shoulder joint ?

i. The **coracohumeral ligament** is a broad band extending from the lateral part of the root of the coracoid process to the greater tubercle. The upper anterior border of this band is free and overlaps the capsule underneath it. It may be reinforced by fibres of the tendon of pectoralis minor which have crossed over the coracoid process through the medial attachment of the coracoacromial ligament.

ii. The **glenohumeral ligaments** are seen convincingly only from within the capsule. They are thought to be important, however, for the stability of the joint. The **superior** glenohumeral ligament arises from the base of the coracoid process where the apex of the labrum joins the long head of the biceps and inserts into the anatomical neck opposite the upper part of the lesser tubercle. There may be a small impression at this insertion called the fovea humeri. The **middle** glenohumeral ligament arises from the upper two thirds of the anterior margin of the glenoid and inserts into the anatomical neck opposite the lower part of the lesser tubercle. The **inferior** glenohumeral ligament arises from the lower anterior and inferior margins of the glenoid and inserts into the anatomical neck below and behind the middle ligament, strengthening the upper anterior part of the axillary pouch. This ligament may not be well developed.

iii. The **transverse humeral ligament** is a broad band passing across the intertubercular sulcus (bicipital groove) between the greater and lesser tubercles, forming the roof of a tunnel for the long head of biceps and its synovial sheath.

iv. A thin fibrous extension of the tendon of insertion of pectoralis major passes up to the capsule.

In spite of these ligamentous bands it is possible to distract the articular surfaces a couple of centimetres by a firm pull on the relaxed arm.

166. What openings are there in the capsule of the shoulder joint ?

i. The long head of biceps passes in its synovial sheath through the tunnel formed by the intertubercular sulcus, greater and lesser tubercles and the transverse humeral ligament.

ii. There is usually an opening for communication between the joint cavity and the subscapularis bursa, usually between the superior and middle glenohumeral ligaments. There may be two openings, one on either side of the middle ligament. There may be no opening.

iii. There is occasionally a posterior opening for communication with the bursa deep to infraspinatus.

167. What movements take place at the shoulder joint ?

Movements of the arm at the shoulder are sometimes described in relation to the plane of the scapula, on a glenoid surface facing forwards as well as laterally, some 30–40° to the coronal plane. In clinical practice, however, the three planes of the body in the anatomical position are used. In the sagittal plane there is 0–180° of forward **flexion**, and 60° of backward **extension**. In the coronal plane there is 0–180° of **abduction**. The opposite movement, of 0–60° of **adduction**, takes the arm across the body, in front or behind, as near the coronal plane as possible. The transverse plane is also sometimes used, the arm starting at 90° of abduction. There is then said to be 0–135° of horizontal flexion and 0–45° of horizontal extension. **Rotation** movements of the arm, in the longitudinal axis of the humerus, are described as medial (internal) or lateral (external), the range varying with the position of the arm. From the anatomical position there is about 70° of lateral rotation and about 100° of medial rotation. With the arm abducted to 90° there is only about 60° of medial rotation.

Only 120° of abduction are possible at the glenohumeral joint. Beyond this angle there would be impaction of the neck of the humerus against the acromion and coracoacromial ligament. As the humerus abducts the intertubercular tunnel ascends the tendon of the long head of the biceps towards its origin from the supraglenoid tubercle. As it does so the humerus laterally rotates so that at the same time the greater tubercle avoids impaction against the coracoacromial ligament. If the humerus is kept in full medial rotation only about 60° of glenohumeral abduction is possible, the greater tubercle impinging against the coracoacromial ligament. If abduction is forced without lateral rotation impingement injuries are likely. There may be crushing of the rotator cuff, fracture of the greater tubercle, or dislocation of the joint. The remaining 60° of abduction occurs between the scapula and the chest wall. Over the whole range of abduction, therefore, the scapula rotates on the chest wall 1° for each 2° of glenohumeral movement. However, the first 30° or so of abduction of the shoulder occur at the glenohumeral joint only. Thereafter a smooth combined movement occurs, the glenohumeral joint abducting its remaining 90°, and the scapula its 60°, a final ratio of 3:2.

The close-packed position of the joint, that is of maximum congruity and tautness of its capsule and ligaments, is in abduction, extension, and lateral rotation, as when, for instance, the arm is about to throw a javelin.

168. What factors contribute to the stability of the shoulder joint ?

i. The head of the humerus is a third of a sphere and articulates with a shallow glenoid surface a third of its size. The glenohumeral joint thus contrasts with the hip joint in that no significant contribution is made to its stability by the contour of the articular surfaces.

ii. The **tendons** of insertion of subscapularis, supraspinatus, infraspinatus and teres minor muscles merge together with the underlying capsule to form a dynamic capsulotendinous envelope or rotator cuff over the anterior, superior and posterior aspects of the joint. When the cross sectional areas of the rotator cuff muscles are added together they are about equal to that of the deltoid, and therefore comprise the most powerful mechanism for stabilising the joint. It is thought that inco-ordination between the components of the cuff may be the cause of recurrent and habitual dislocations in some patients. The imbalance is mostly in favour of the lateral rotators, causing anterior dislocation. The tendon of the long head of biceps, passing within the joint from the supraglenoid tubercle and posterosuperior part of the labrum through the intertubercular tunnel, especially in neutral to lateral rotation, resists upward subluxation of the head of the humerus. The tendon of the long head of triceps, arising from the infraglenoid tubercle, resists downward subluxation of the abducted shoulder.

iii. The part played by the **ligaments** of the joint is difficult to evaluate. The coracohumeral and superior glenohumeral ligaments help prevent downward subluxation of the head of the humerus, but can be easily stretched over a week or two if the deltoid and supraspinatus muscles are inactive. Experiments on cadaver shoulders have led to the conclusion that, in the anatomical position of the shoulder and up to 45° of abduction, anterior stability is provided mainly by subscapularis. As abduction reaches 45° the anterior stability provided by subscapularis is reinforced by tension in the middle glenohumeral ligament and the anterosuperior fibres of the inferior glenohumeral ligament. As abduction reaches 90° the inferior glenohumeral ligament plays its part by resisting anterior subluxation as the head of the humerus rotates laterally. It is thought that absence or lack of development of the lower ligaments may contribute to the ease of a first anterior and then recurrent anterior dislocation in some patients.

iv. The **coracoacromial arch**, formed by the acromion, coracoacromial ligament and the coracoid process, and separated from the rotator cuff by the subacromial bursa, acts as a more remote stabilising mechanism. However, the acromion particularly becomes a factor in stability when the rotator cuff and tendon of the long head of biceps are degenerate and torn, as commonly is the case in the aged. On contraction of the deltoid the head of the humerus rises to impinge against the acromion.

169. What muscles abduct the humerus, and where are they inserted ?

Abduction of the humerus is brought about by the action of —

i. the **deltoid — short rotator couple**. The tendon of the deltoid muscle inserts into the V-shaped deltoid tuberosity a little above the middle of the lateral side of the shaft. The anterior and posterior parallel fibres of the muscle reach the limbs of the V, and the middle multipennate fibres from the acromion insert into the central area. The tendon of subscapularis inserts into the lesser tubercle. The tendon of teres minor inserts into the lower facet of the greater tubercle. Acting alone the deltoid causes upward subluxation of the head. The short rotators, excluding supraspinatus, acting alone depress the head, and can even cause subluxation. Acting together, the deltoid and the lower short rotators form a powerful force-couple to abduct the arm.

ii. **supraspinatus**, the tendon of which inserts into the uppermost facet of the greater tubercle. It acts with the deltoid throughout the whole range of abduction. It is not essential for abduction, however, since this is still possible even when supraspinatus is paralysed or its tendon is torn, but power cannot be sustained normally.

iii. **biceps brachii**, which is not attached to the humerus. In the anatomical position the tendon of the long head is not effective in abduction, but in 90° of lateral rotation the tendon forms a straight line with the belly of the muscle and can then make a significant contribution. Biceps may be used in a trick movement to abduct the arm when the deltoid muscle is paralysed.

170. Where are the muscles that medially rotate the humerus attached ?

The tendon of **pectoralis major** inserts into the lateral lip of the intertubercular sulcus, down to the deltoid tuberosity. The clavicular fibres insert via the anterior lamina of the tendon, the manubrial and upper sternocostal fibres form the intermediate lamina and the lower sternocostal fibres form the recurved posterior lamina, so that the lowest muscle fibres pass posteriorly to the uppermost part of the tendon. In this way the round contour of the anterior axillary fold is formed. The tendon of **latissimus dorsi** inserts into the floor of the intertubercular sulcus in front of the tendon of **teres major**. The tendon of teres major inserts into the medial lip of the intertubercular sulcus.

The tendon of **subscapularis** inserts into the lesser tubercle. The tendinous fibres of the anterior part of the **deltoid** insert into the anterior part of the deltoid tuberosity.

171. Where are the muscles that laterally rotate the humerus attached ?

The tendon of **infraspinatus** inserts into the middle facet of the greater tubercle.

The tendon of **teres minor** inserts into the lower facet of the greater tubercle.

The muscles of the rotator cuff are said to contribute 80% of the power of external rotation.

The posterior fibres **deltoid** insert into the posterior part of the deltoid tuberosity.

172. What muscles adduct the humerus ?

The adductor muscles of the shoulder constitute a powerful group. They are —
  pectoralis major, inserted into the lateral lip of the intertubercular sulcus.

latissimus dorsi, inserted into the floor of the intertubercular sulcus.
teres major, inserted into the medial lip of the intertubercular sulcus.
teres minor, inserted into the lowest facet on the greater tubercle.
subscapularis, inserted into the lesser tubercle.

173. What muscles flex the humerus ?

The flexor muscles of the shoulder are, in contrast to the adductors, relatively weak. They comprise —
coracobrachialis, inserted into a rough strip along the medial border of the middle of the shaft, limited below by a nutrient foramen.
anterior fibres of **deltoid**, inserted into the anterior margin of the deltoid tuberosity in line with the lateral lip of the intertubercular sulcus.
clavicular fibres of **pectoralis major**, when the arm starts from a position of extension, inserted into the lateral lip of the intertubercular sulcus.
biceps brachii, acting indirectly, inserted into the radial tuberosity.

174. What muscles extend the humerus ?

The extensor group of muscles comprise —
sternocostal fibres of **pectoralis major**, when the arm starts from a position of flexion, inserted into the lateral lip of the intertubercular sulcus.
latissimus dorsi, inserted by a flat tendon into the floor of the intertubercular sulcus.
teres major, inserted into the medial lip of the intertubercular sulcus.
posterior fibres of **deltoid**, inserted into the posterior margin of the deltoid tuberosity.

175. What is the sensory nerve supply to the shoulder joint ?

The nerve supply of the shoulder joint illustrates the general rule (Hilton's law), that the nerves which supply muscles acting over a joint also supply the capsule of the joint and the skin over it. Branches of the **axillary nerve** reach the anterior and inferior parts of the capsule. Other branches to this part of the joint come from the **posterior cord** or **radial nerve**. Branches of the **suprascapular nerve** reach the superior and posterior parts of the capsule. Branches of the **lateral pectoral nerve** reach the anterior and superior parts of the capsule.

176. What is the blood supply of the rotator cuff ?

The tendinous part of the rotator cuff is well vascularised by anastomoses between osseous and muscular vessels. A branch of the anterior circumflex humeral artery ascends the intertubercular sulcus, and penetrates the head of the humerus. Within the bone it gives branches which enter the rotator cuff via the tendinous insertions. Before this branch of the anterior circumflex humeral artery penetrates the bone it gives a superficial branch which also supplies the tendinous cuff. Branches of the suprascapular and subscapular arteries anastomose and prolong a network beyond the musculotendinous junction into the tendinous cuff. The area of anastomosis between the osseous and muscular vessels is known as the critical zone and corresponds to the site of frequent tendinous pathology. The subdeltoid bursa is supplied by underlying vessels.

177. What bursae are there around the shoulder joint ?

There are seven.

i. The **subdeltoid** or subacromial bursa separates the middle fibres of the deltoid, the acromion and coracoacromial ligament from the upper surfaces of supraspinatus, infraspinatus and subscapularis muscles and tendons. Of varying development, it may be in the form of two separate bursae, subacromial and subdeltoid.

ii. The **subscapular** bursa separates the upper, more tendinous part of the subscapularis from the neck of the scapula. It communicates commonly with the synovial cavity of the shoulder joint through one or two openings in the capsule.

iii. The **subcoracoid** bursa separates the upper edge of subscapularis from the smooth undersurface of the coracoid process. It may communicate with the subscapular bursa.

iv. The **infraspinatus** bursa separates the infraspinatus from the neck of the scapula, and may communicate with the joint cavity.

v. **Other** bursae are found over the subcutaneous surface of the acromion, between the long head of triceps and teres major, and between teres major and latissimus dorsi.

178. What is the surgical neck of the humerus ?

The surgical neck of the humerus is the horizontal plane of bone below the head and tubercles through which fractures frequently occur. It is in contrast to the anatomical neck which encircles the articular surface of the head.

179. How does the upper end of the humerus develop ?

The primary centre of ossification of the shaft appears at about the 8th week of fetal life. The centre of ossification of the head of the humerus may be present at birth, or else develops during the first 6 months. The centre for the greater tubercle appears about the 2nd year. The centre for the lesser tubercle appears about the 5th year, but may be difficult to see on standard X-ray films. The upper epiphyses fuse together at about 7 years. The resulting epiphyseal plate is cone-shaped, the apex of which is behind and medial to the centre of the head. The lateral margin of the epiphyseal plate is at the surgical neck and is extracapsular. The medial margin reaches the joint surface 2–3 mm above the articular margin so that a small portion of the metaphysis is intracapsular, a point of importance in the spread of infection into the joint in osteomyelitis. The upper epiphysis contributes 80% of the growth of the humerus. During growth the perichondrium is stronger on the posterior and medial surfaces than the anterior and lateral surfaces, a fact of importance in the anatomy of injury of the part.

180. Where do the flexor muscles of the elbow arise from the humerus ?

Brachialis arises from the anterior aspect of the lower half to two thirds of the bone, its upper origin embracing the distal part of the deltoid tuberosity. More medially it reaches between the deltoid tuberosity and the insertion of coracobrachialis. **Brachioradialis** arises from the anterior surface of the upper two thirds of the lateral supracondylar ridge. Although **biceps** brachii does not normally arise from the humerus, about one in five arms show an accessory humeral head of origin of biceps from above the origin of brachialis.

181. Where does the extensor muscle of the elbow arise from the humerus ?

The lateral and medial heads of **triceps** arise from the posterior aspect of the bone. The lateral head has an oblique linear aponeurotic origin from above the spiral groove. The medial head has a more extensive origin from the whole surface from below the spiral groove to just above the olecranon fossa. It also arises from both intermuscular septa. The long head does not arise from the humerus, but from the infraglenoid tubercle of the scapula.

182. What muscles arise from the lateral epicondyle ?

Anteriorly, from the lateral epicondyle, arise extensor carpi radialis brevis, extensor digitorum longus, extensor digiti minimi, and extensor carpi ulnaris. Laterally arises supinator and posteriorly anconeus.

183. What muscles arise from the medial epicondyle ?

Arising from the medial epicondyle anteriorly are pronator teres, flexor carpi radialis, palmaris longus, flexor digitorum superficialis, and flexor carpi ulnaris.

184. What is a supracondylar spur ?

It is a spur of bone, arising from the anteromedial aspect of the supra-condylar surface, about 5 cm above the medial epicondyle. It is said to be present in 1% of bodies. The ligament of Struthers continues from its tip to join the medial epicondyle, forming a fibrosseous passage between it and the bone through which passes the median nerve, and often the brachial artery. In high division of the brachial artery the ulnar artery may join the median nerve under the ligament. The ligament may ossify to form a supracondylar foramen. This unusual anatomy may be the site of entrapment of the median nerve.

185. What nerves are in contact with the humerus ?

i. The **axillary nerve** arises from the posterior cord and passes backwards through the quadrangular space bounded laterally by the surgical neck of the bone, medially by the long head of triceps, superiorly by the subscapularis, capsule of the shoulder joint and teres minor, and inferiorly by the teres major. After giving off the posterior branch it continues as its anterior branch round the posterior and lateral aspects of the surgical neck, supplying the deltoid, and a few branches to the skin. It is liable to injury following anterior dislocation of the shoulder; the reported incidence of isolated axillary nerve damage varies from 10–20%.

ii. The **radial nerve** lies against the bone in the spiral groove, or is separated from it by the uppermost fibres of the medial head of triceps. Branches of the radial nerve to the lateral and medial heads of triceps arise from the main trunk in the spiral groove, often high up, and continue at first as collaterals. Thus damage to the radial nerve in the spiral groove following a fracture of the shaft may spare the more proximal innervation of the triceps. The posterior cutaneous nerve of the forearm may also arise high up in the groove and accompany the parent nerve. The lower lateral cutaneous nerve arises in the groove. The radial nerve reaches the lateral border of the bone, where it pierces the lateral intermuscular septum, just below the insertion of the deltoid. Here it is relatively superficial and fixed, and susceptible to direct injury.

iii. The **musculocutaneous nerve** may touch the bone after it emerges from coracobrachialis in subjects in whom the origin of brachialis does not reach as high as usual.

iv. The **ulnar nerve** lies against the back of the medial epicondyle as it approaches the cubital tunnel, formed by the medial collateral ligament of the elbow joint, the olecranon, and overlying fascia.

v. **The posterior cutaneous nerve of the forearm** descends behind the lateral epicondyle.

186. How does the lower end of the humerus ossify ?

The lower end of the humerus, the condyle, is entirely cartilaginous at birth. The ossific centre for the lateral part of the cartilaginous condyle appears during the **1st** year and gives rise to the capitulum and the lateral part of the trochlea. The centre for the medial epicondyle appears next, at about **5–6** years. It is liable to avulsion injury. The centre for the medial part of the condyle, which gives rise to the medial part of the trochlea, appears at about 9 years. The centre for the lateral epicondyle appears as a thin flake of bone at about 12 years. The centres for the lateral epicondyle, capitulum and trochlea fuse together at about 15 years, and with the shaft and medial epicondyle about 2 years later. In general the centres of ossification appear at an earlier date in girls than boys. The centres of ossification need to be distinguished from fractures when looking at X-ray films of young injured elbows.

187. Where is the capsule of the elbow joint attached to the humerus ?

Anteriorly the capsule attaches to the medial margin of the trochlea, to immediately above the coronoid and radial fossae, and to the lower edge of the lateral epicondyle. **Posteriorly** the capsule attaches to the lower edge of the lateral epicondyle, to the sides and the upper border of the olecranon fossa, and to the medial margin of the trochlea.

188. What are the fat pads of the elbow joint ?

Two accumulations of fat are present under the capsule of the elbow joint, between it and the synovial membrane. An anterior one overlies the upper part of the coronoid fossa. Because the coronoid fossa is shallow this fat pad projects in front of the margins of the fossa and can be seen on a lateral X-ray film of the normal joint as a triangular translucency. The posterior fat pad overlies the olecranon fossa at a slightly lower level. Because this fossa is deep, the posterior fat pad is contained within it and cannot be seen on the usual lateral X-ray film, with the elbow flexed to a right angle. There are also thin layers of fat overlying the capsule in front and behind, between it and the adjacent muscles. If the elbow is extended, however, the posterior fat pad is displaced by the olecranon and then becomes visible. When an injury causes an effusion into the joint the fat pads are displaced. On the lateral X-ray film with the elbow bent to a right angle the anterior pad is pushed forward and the posterior pad becomes visible. These X-ray film changes are known as the **fat pad signs**.

THE RADIUS AND ULNA

189. Where is the capsule of the elbow joint attached distally?

The anterior part of the capsule is attached to the anterior part of the anular ligament and to the coronoid process of the ulna near its articular margin. Medially the capsule is reinforced to form the medial ligament attached around the sublime tubercle of the coronoid process and to the medial side of the olecranon. Between these attachments of the medial ligament, that of the capsule to close to the articular margin of the trochlear notch appears somewhat deficient. Posteriorly the capsule is attached near to the articular margins of the superior surface and sides of the olecranon and the posterior part of the anular ligament. Laterally the capsule is reinforced to form the lateral ligament which blends with the lateral part of the anular ligament. Inferiorly the capsule between the medial side of the neck of the radius and the inferior margin of the radial notch of the ulna is thinly reinforced to form the quadrate ligament. The capsule between the lower margin of the anular ligament and the neck of the radius below the articular surface of the head is only a short, loose and thin membrane.

190.   What are the ligaments of the elbow joint ?

i. The **anular ligament** is part of both the elbow and superior radio-ulnar joints. It is attached anteriorly to the anterior margin of the radial notch of the ulna. Posteriorly its attachment to the posterior margin of the radial notch is extended, reaching to the upper part of the supinator crest. It forms a strong band around the head of the radius. The lower diameter of the osseoligamentous ring so formed is narrower than the upper, thus presenting a slightly cup-shaped restraint to distal distraction of the head of the radius. Its superficial surface is blended with the lateral ligament. The upper part of the deep surface of the ligament is covered by fibrocartilage, the lower part by synovial membrane. The latter bulges down a little and is covered by only a thin layer of fibrous tissue extending from the lower margin of the ligament to the neck of the radius just beyond its articular surface.

ii. The **lateral ligament** arises from a roughened area of the lateral epicondyle of the humerus just below its most prominent part. It radiates distally to blend with the anular ligament anteriorly, laterally and posteriorly. It reaches the coronoid process anteriorly, and the olecranon posteriorly, where it is especially strong.

iii. The **medial ligament** radiates distally from the medial epicondyle of the humerus. It is multifascicular. The anterior part is prominent, wide and strong, and extends to the medial proximal area of the coronoid process, around the medial or sublime tubercle. In it can be distinguished an anterior band which is tight in extension, and a posterior band which is tight in flexion. There is an oblique band of fibres which is attached only to the ulna, extending from the sublime tubercle to the medial side of the base of the olecranon near the articular margin. A little fat or synovium often protrudes from beneath the oblique band, where the capsular attachment is weak. Intermediate fascicles of the medial ligament mingle with the oblique band. The fascicles of the posterior part of the medial ligament are attached to the side of the olecranon near the articular margin behind and mingle with the oblique band below. They are relaxed in extension but tight in flexion. The bands of the medial ligament form a groove for the ulnar nerve.

iv. The **quadrate ligament** is the thickening of the capsule which extends from just below the lower border of the radial notch of the ulna to the neck of the radius.

v. The **anterior capsule** is as strong as a ligament. It has a deep layer of a few thin transversely running fibres, a main intermediate layer of strong vertical fibres, and a superficial layer of one or two oblique bands passing downwards and laterally to the anular ligament. In children the anterior capsule and the collateral ligaments are stronger than the bone. The anterior capsule is thought to cause the olecranon to act as a fulcrum, fracturing the distal humerus, when excess extension strain is exerted on the ulna. In contrast the posterior capsule is thin and is easily displaced by an effusion within the elbow joint.

191. What factors contribute to the stability of the elbow joint ?

The coronoid process, trochlear notch and olecranon of the ulna fit well into the coronoid fossa, trochlea and olecranon fossa of the humerus. The radial head is held tightly to the ulna by the anular ligament, and secondarily by the lateral ligament to the lateral epicondyle of the humerus. The medial ligament, especially its anterior part, is strong. The elbow is most stable, and least liable to dislocation, in flexion. In this position the posterior fibres of the anterior part of the medial ligament are very taut, and the medial side of the trochlear notch is supported by the larger surface of the medial flange of the trochlea. Conversely, the joint is most likely to dislocate in extension.

192. What are the surface features of the elbow joint ?

Bony projections are the medial and lateral epicondyles, and the olecranon. The head of the radius is palpable, especially on pronation and supination. In extension the epicondyles and the point of the olecranon are on the same horizontal level. In flexion the olecranon descends to outline with the epicondyles an equilateral triangle. In a supracondylar fracture the triangle remains intact. In a dislocation of the joint the triangle is distorted. Posterolaterally the normal joint shows a slight hollow between the olecranon and the lateral epicondyle. In the lower part of the floor of the hollow is the head of the radius covered by anconeus. The hollow is lost if an effusion develops within the joint. It is a good site for aspiration of the joint or the injection of local anaesthetic into it. The ulnar nerve is easily palpated in the groove between the back of the medial epicondyle and the olecranon.

193. What features of a lateral X-ray film of an elbow confirm the normal alignment of the joint ?

The joint is usually viewed flexed. The line of the shaft of the radius, extended proximally through the neck and head, normally passes centrally through the capitulum. It is easy to miss a subluxation or dislocation of the head of the radius if this position is not checked. The curve of the line of the anterior surface of the coronoid process should continue smoothly upwards onto the anterior surface of the lower humerus. It would not do so if the elbow joint were dislocated. An extension downwards of the line of the anterior surface of the lower shaft of the humerus normally passes through the middle of the trochlea. It would not do so if the lower end of the humerus were fractured and displaced.

194. Where are the muscles that extend the elbow joint attached to the forearm bones ?

i. The tendon of **triceps**, the prime mover of extension, is inserted into the posterior part of the upper surface of the olecranon. It is often separated from the capsular attachment more anteriorly by a bursa, in contact with the bone.

ii. **Anconeus**, arising by tendinous fibres from the posterior aspect of the lateral epicondyle of the humerus, is inserted into the lateral aspect of the olecranon posterior to the attachment of the anular ligament and lateral ligament, and into the upper quarter of the extensor surface of the shaft between the supinator ridge and the subcutaneous border. Its power of extension is weak. Anconeus is also active during both pronation and supination, and is regarded as having mainly a joint stabilising rather than prime moving function.

195. Where are the muscles that flex the elbow joint attached to the forearm bones ?

i. **Brachialis** is inserted into the triangular rough area of the coronoid process. It is active in all movements of the joint and in any position of the forearm. It is described as the 'workhorse' among elbow flexors.

ii. **Biceps brachii** is inserted into the posterior part of the tuberosity of the radius, and by the bicipital aponeurosis into the upper subcutaneous border of the ulna. It is especially active in rapid and forced movements, and is a powerful supinator in flexion.

iii. **Brachioradialis** is inserted into the base of the styloid process of the radius. It comes into action when force and speed of contraction are needed. It tends to restore the pronated or supinated forearm to neutral.

iv. Muscles arising from the common flexor origin assist in flexion of the elbow. They comprise pronator teres, flexor carpi radialis, flexor digitorum superficialis, palmaris longus and flexor carpi ulnaris. Only the **pronator teres** is inserted into a forearm bone, into the lateral aspect of the maximum convexity of the radius, between the anterior and posterior oblique lines.

196. What is the nerve supply of the elbow joint ?

The elbow joint receives branches from all four main nerves in the arm. The anterior part of the capsule is innervated by a branch of the branch of the musculocutaneous nerve to brachialis, which descends along the medial side of the brachialis, and by a recurrent branch of the median nerve which arises just before it passes between the two heads of the pronator teres. The posteromedial part of the capsule receives one or more fine branches from the ulnar nerve as it passes behind the medial epicondyle. The anterolateral, lateral and posterolateral parts of the capsule receive several branches from the radial nerve.

197. What bursae are there around the elbow joint ?

i. The most constant **subcutaneous** bursa is that covering the posterior surface of the olecranon. Subcutaneous bursae over the medial and lateral epicondyles occur.

ii. Constant **deep** bursae include one between the tendon of biceps and the anterior part of the tuberosity of the radius; one between the tendon of biceps in front, and the flexor digitorum profundus and supinator behind, the tuberosity of the radius; and one between the insertion of the tendon of triceps and the anterior part of the upper surface of the olecranon immediately in front of it. Other deep bursae described include one within the tendon of triceps near its insertion, one between the origin of anconeus and the capsule over the head of radius, and one between the common extensor origin and supinator.

198. What are some clinically relevant features of the superior radio-ulnar joint ?

i. The head of the radius is part both of the elbow joint and the superior radio-ulnar joint. Any malalignment of the surfaces of the joint, for example after injury, may severely limit movements of the elbow and forearm.

ii. The overhang of the head on the neck of the radius is greatest anteriorly and laterally. In supination, therefore, the maximum degree of overhang is contained by the anular ligament. In pronation the lateral part of the neck becomes anterior, and the anterior part becomes medial against the radial notch of the ulna, so that there is a less efficient grip of the head by the anular ligament.

iii. The articular surface of the capitulum does not extend posteriorly and inferiorly beyond the mid-coronal plane of the joint. Therefore, in full or slight hyper-extension, only a reduced amount of the surface of the radial head is in contact with the articular surface of the capitulum, and the joint is least stable.

iv. The strength of the attachment of the anular ligament to the neck of the radius is particularly weak up to the age of about 5 years.

'Pulled elbow' is a condition caused by a sudden distraction force applied to the forearm, where there follows a partial slipping of the inferior margin of the anular ligament over the radial head, with sometimes a tear of the thin capsular attachment of the ligament to the neck of the radius. A meniscoid fold of synovium may become interposed between the radial and capitular articular surfaces, and may be the cause of the inhibition of elbow and forearm movements that follows. As would be expected from the above features the injury occurs usually in a child under 5 years, when the forearm is in pronation and the elbow is extended.

199. How can the rotational position of the forearm be assessed on an anteroposterior view X-ray film of the elbow ?

The rotational position of the radius can be assessed by examination of the outline of its tuberosity. The tuberosity is most evident in full supination. As the radius pronates the medial prominence of the tuberosity fades and a cyst-like outline of the en-face view of the tuberosity appears in the bone, which moves to contact the lateral cortex in full pronation.

200. What are the features of the inferior radio-ulnar joint ?

i. The articular surface of the ulna is a convex crescent which covers about two thirds of the surface of the head. The points of the crescent reach to the base of the styloid process.

ii. The distal articular surface is formed by the ulnar notch of the radius and the concave proximal surface of the triangular fibrocartilage.

iii. The triangular fibrocartilage is attached by its thinner base to the distal border of the ulnar notch of the radius, and by its thicker apex to a depression between the styloid process and the articular surface of the head of the ulna. It is concave on both proximal and distal surfaces, with thickened borders. It undergoes degenerative changes with age, half the fibrocartilages examined in subjects over 60 years showing perforations, with 'kissing' areas of damage of the opposing articular surfaces.

iv. The capsule forms loose sac-like membranes anteriorly and posteriorly. Superiorly the capsule is very thin allowing a protrusion of synovium, the saccus sacciformis, to form in front of the distal interosseous membrane, deep to pronator quadratus.

v. The position of maximal congruity of the joint, and therefore of stability, is when the greatest vertical heights of the opposing articular surfaces of the ulnar notch of the radius and the convex crescent of the head of the ulna are in apposition. It is near to the mid-position of pronation and supination. This is the position when the triangular fibrocartilage is most tense. Beyond this position in either direction there is more limited contact between the articular surfaces and therefore reduced stability.

201. What are the features of the interosseous membrane ?

The interosseous membrane is an aponeurotic sheet stretched between the interosseous borders of the forearm bones, from just below the tuberosity of the radius to the inferior radio-ulnar joint. Distally the radial attachment of the membrane follows the posterior border of the triangular surface above the ulnar notch. About halfway down the forearm the membrane is thickened to form a band about 2.5 cm wide which is directed obliquely downwards from the radius to the ulna. The thickest part of the membrane is slightly proximal to the midpoint of the radius, and is about twice the thickness of the membrane above and below it. It has been found that this central band contributes nearly three quarters of the longitudinal resistance of the membrane to the proximal migration of the radius after the head of the radius has been excised. The interosseous membrane is also found to be under maximal tension in the prone position and to relax progressively during supination. The upper edge of the membrane corresponds to the upper border of abductor poilicis longus, over which pass the posterior interosseous vessels between it and the lowest fibres of the supinator. The anterior surface of the membrane gives origin to fibres of the flexor pollicis longus, part of flexor digitorum profundus, and pronator quadratus, and is in contact with the anterior interosseous vessels and nerve. The posterior surface gives origin to fibres of the abductor pollicis longus, extensor pollicis longus, extensor pollicis brevis and extensor indicis. The posterior interosseous nerve continues to the dorsum of the carpus in contact with the membrane distal to the abductor pollicis longus. The posterior interosseous vessels are reinforced distally by the anterior interosseous vessels which pierce the membrane from in front.

202. What are pronation and supination ?

In the unnatural 'anatomical position', the palm of the hand faces forwards and the thumb points laterally. The forearm is then in full supination. If the elbow is flexed from the anatomical position of full extension to 90° the palm then faces upwards. The position of full pronation follows, in this flexed position of the elbow, when the hand is turned over so that the palm faces downwards and the thumb points medially. The mid- or zero position of pronation or supination is then the one in which the palm faces medially and the thumb points superiorly. There is normally 90° of supination and 85° of pronation. In relation to the bones of the forearm the constant anatomical axis of rotation passes between the centre of the head of the radius and the attachment of the apex of the triangular fibrocartilage to the the ulna. In relation to the hand the functional axis of rotation can be made to pass through any finger by slight secondary movements of the head of the ulna. This adjustment is seen most when the axis of rotation passes through the thumb. In the movement from neutral to pronation the head of the ulna is carried slightly anteriorly and laterally, and from neutral to supination the head of the ulna is carried anteriorly and medially. The slight movements of the ulna which accompany pronation and supination are achieved by active extension and flexion, and passive abduction and adduction of the bone at the elbow. The muscles that supinate the forearm are stronger than those that pronate.

203. Where are the muscles attached that pronate the forearm ?

i. **Pronator quadratus** is the prime pronator. It has a linear origin from the curved pronator crest of the distal quarter of the ulna and is inserted into the anterior surface of the distal quarter of the radius. It is sometimes laminated, with two or more layers, the fasciculi running in different directions. Fibres can continue downwards onto the carpus on either the radial or ulnar sides.

ii. **Pronator teres** is recruited as a reinforcement to pronation whenever the action is rapid or against resistance. It has two heads. The **superficial head** arises from the humerus just proximal to the medial epicondyle (and from the ligament of Struthers when this is present), from the front of the medial epicondyle in the common flexor origin, from the septum between it and flexor carpi radialis, and from the overlying fascia which is continuous with the bicipital aponeurosis. The latter connection is significant in that the tendinous origin of the muscle and the upper edge of the aponeurosis can form the proximal of three fibrous arches under which the median nerve travels to the deep surface of the flexor digitorum superficialis. A variable **deep head** arises from the coronoid process, usually from its medial border just distal to the sublime tubercle. It is absent in 6% of subjects. It may be rudimentary. It may arise elsewhere from the coronoid process, or from the interosseous membrane. The two heads unite to form the second or intermediate arch for the median nerve to pass under. Either limb of this second arch can present a rather sharp fibrous edge. The muscle forms a flattened glistening tendon, a useful landmark when dissecting in the area, inserted into the lateral aspect of the shaft of the radius, onto a slightly roughened surface between the anterior and posterior oblique lines at the site of maximum lateral convexity of the bone. Each of the two fibrous arches described can be the site of entrapment neuropathy of the median nerve.

204. Where are the muscles attached that supinate the forearm ?

i. **Supinator** has a superficial and a deep component. **Superficial fibres** arise from the distal margin of the lateral epicondyle, the lateral ligament of the elbow joint, and the posterior part of the anular ligament. They pass downwards to the upper part of the anterior oblique line of the radius. The proximal border of the superficial head presents a tendinous, thickened edge. This can take the form of a gentle curve to its attachment just below the tuberosity of the radius, or it can recurve and form an additional attachment by an inverted arch to the medial aspect of the lateral epicondyle just lateral to the capitulum. If the thickening is prominent it is described as the **arcade of Frohse**, important because under it passes the posterior interosseous nerve between the superficial and deep laminae of the muscle. The arcade of Frohse can be the site of entrapment of the posterior interosseous nerve. The radial recurrent artery runs across it. The **deep** fibres of the muscle arise from the supinator crest of the ulna and from a thin zone of the concavity medial to it. They wrap round the shaft of the radius transversely deep to the superficial fibres to insert into the bone between the upper parts of the oblique lines, down to the insertion of the pronator teres.

ii. The short head of **biceps** arises in common with the coracobrachialis from the tip of the coracoid process of the scapula. The long head arises from the supraglenoid tubercle of the scapula and the glenoid labrum. The common tendon formed low down inserts by a flat tendon, in the sagittal plane into the posterior part of the tuberosity of the radius, separated from the anterior part of the tuberosity by a bursa. The bicipital aponeurosis spreads medially into the deep fascia of the forearm to reach the upper subcutaneous border of the ulna and is an accessory insertion for flexion of the elbow.

205. From where do the muscles arise that primarily flex the wrist ?

Flexor carpi radialis and **palmaris longus** arise from the common flexor origin from the anterior surface of the medial epicondyle of the humerus. They do not have origin from the radius or ulna. **Flexor carpi ulnaris** has two heads of origin: the humeral head arises from the common flexor origin; the ulnar head arises from the lower part of the medial margin of the olecranon and the proximal two thirds of the posterior, subcutaneous border of the ulna, by an aponeurosis in common with that of the extensor carpi ulnaris and flexor digitorum profundus.

206. What is the cubital tunnel ?

Cubital tunnel is the clinical term applied to the path taken by the ulnar nerve from behind the medial epicondyle to between the two heads of flexor carpi ulnaris and into the flexor-pronator musculature of the forearm. The floor of the tunnel, as seen at operation, comprises its anterior and lateral boundaries, which are the distal part of the medial intermuscular septum and the medial head of triceps, the medial epicondyle, the medial ligament of the elbow joint and the most proximal fibres of origin of flexor digitorum profundus. The roof of the tunnel is the fascia covering its posterior and medial boundaries, and distally the gap between the two heads of flexor carpi ulnaris. The fascial roof between the two heads of flexor carpi ulnaris is thickened, forming Osborne's band, between the medial epicondyle and the olecranon. The epitrochleoanconeus, present in one in five bodies, is a small muscle which arises from the back of the medial epicondyle and is inserted into the olecranon, roofing over part of the cubital tunnel. The significance of the tunnel is that it is a site of ulnar nerve entrapment. The diameter of the tunnel is reduced by half, from about 6 mm to 3 mm, from full extension to full flexion of the elbow, and the pressure within it increases by about 50%. The pressure within the tunnel can be increased further if then, with the elbow flexed, the wrist is extended, and even further again if, with the elbow flexed and the wrist extended, the arm is elevated above the head. The fasciculi of the ulnar nerve are arranged such that if the nerve is compressed sufficiently by a tight fascial band the superficial motor and sensory axons to the hand may be affected. If the nerve is compressed on its deeper aspect by a bulge of the capsule of the elbow joint or medial head of triceps the deep motor axons influencing grip strength and wrist flexion may be affected.

207. From where do the muscles arise that primarily extend the wrist ?

Extensor carpi radialis longus arises from the lower third of the lateral supracondylar line and the adjacent intermuscular septum.

Extensor carpi radialis brevis arises from the anterior surface of the lateral epicondyle in the common extensor origin and from the lateral collateral ligament of the elbow joint.

Extensor carpi ulnaris arises from the lateral epicondyle in the common extensor origin and from the posterior, subcutaneous border of the ulna in common with flexor carpi ulnaris.

208. From where do the long or extrinsic muscles arise that flex the fingers ?

Flexor digitorum superficialis has a variably extensive origin. There is a constant origin from the medial epicondyle in the common flexor origin, and from along the anterior band of the medial ligament of the elbow joint to the sublime tubercle of the coronoid process. This may be the only origin of the muscle. Usually there is another origin from the anterior aspect of the upper part of the interosseous membrane, and often the origin extends from the interosseous membrane to the upper part of the anterior oblique line of the radius from just below the tuberosity to the insertion of pronator teres. The gap between the origin of the muscle from the coronoid process and the interosseous membrane can form the third fibrous arch under which pass the median nerve and its anterior interosseous branch. It is another potential site for nerve entrapment. If the anterior interosseous nerve arises laterally from the median nerve it is more vulnerable to compression by the fibrous arch than if it arises posteriorly.

Flexor digitorum profundus arises from the upper three quarters of the medial aspect and the upper two thirds of the anterior aspect of the shaft of the ulna and the adjacent half of the interosseous membrane. It embraces the insertion of brachialis into the coronoid process and reaches down to the attachment of pronator quadratus. It shares a common aponeurosis of origin from the posterior border of the ulna with flexor and extensor carpi ulnares.

209. From where do the muscles arise that extend the fingers ?

Extensor digitorum communis and **extensor digiti minimi** arise from the anterior surface of the lateral epicondyle of the humerus in the common extensor origin.

Extensor indicis arises from the lower third of the lateral part of the posterior surface of the shaft of the ulna outlined by the vertical ridge and the interosseous border, and the adjacent posterior surface of the interosseous membrane, distal to extensor pollicis longus.

210. From where do the long or extrinsic muscles of the thumb arise ?

Extensor pollicis longus arises from the middle third of the lateral part of the posterior surface of the shaft of the ulna, outlined by the vertical ridge and the interosseous border, and from the adjacent posterior surface of the interosseous membrane. Its origin is distal to that of abductor pollicis longus and proximal to that of extensor indicis.

Extensor pollicis brevis arises from the posterior surface of the shaft of the radius distal to the origin of abductor pollicis longus, and from the adjacent posterior surface of the interosseous membrane.

Abductor pollicis longus arises from both bones of the forearm and the intervening interosseous membrane. From the ulna it arises from the upper part of the lateral part of the posterior surface, limited above by supinator from its crest, and below by the origin of extensor pollicis longus. From the radius it arises from the posterior surface of the middle third, above the origin of extensor pollicis brevis.

Flexor pollicis longus arises from the anterior surface of the upper two thirds of the shaft of the radius and the adjacent half of the interosseous membrane, from the level of the tuberosity to the attachment of pronator quadratus. An additional head of origin known as **Gantzer's muscle** is present in about two out of three bodies. It usually arises from the medial epicondyle, from the deepest layer of the common flexor origin, or occasionally from the coronoid process, and inserts into the proximal part of the tendon of flexor pollicis longus. It is of clinical significance since it can be the cause of entrapment of the anterior interosseous nerve. A fibrous band within the muscle may compress the nerve against the flexor digitorum superficialis anterior to it. A fibrous contraction of Gantzer's muscle can cause a flexion contracture of the thumb.

211. Where is the posterior interosseous nerve related to the radius ?

The posterior interosseous nerve passes through supinator around the upper third of the shaft of the bone. To enter the muscle proximally it passes through the zone of the arcade of Frohse (see question 204), and travels between the deep and superficial heads. Its course in supinator is called the supinator tunnel. Entrapment of the nerve in this part can cause symptoms suggestive of tennis elbow syndrome. In about one in four subjects the nerve passes across a bare area of the radius at the level of the tuberosity, where the deep fibres of supinator fail to insert to the same proximal extent as the superficial fibres. The nerve can be injured by fracture or surgery in this area. Damage to the nerve causes paralysis of supinator, extensor digitorum communis, extensor digiti minimi, extensor carpi ulnaris, abductor pollicis longus, extensor pollicis longus, extensor pollicis brevis and extensor indicis. There is no sensory loss.

212. Where are the attachments of the extensor retinaculum ?

The extensor retinaculum is a thickening of the deep fascia of the distal, dorsal part of the forearm, about 2–3 cm wide. Radially it is attached to the sharp lateral border of the distal radius. To the ulnar side the fibres of the retinaculum glide obliquely distally over the head of the ulna to reach the dorsum of the triquetral, and the pisiform, and the ulnar side of the base of the fifth metacarpal. The head of the ulna is thus allowed free movement under the retinaculum in pronation and supination, but at the same time the carpus is held strongly to the radius. Proximal to the main stream of retinacular fibres on the ulnar side, and not attached to the radius, some deeper fibres form a band which attaches to the edges of the groove on the back of the distal ulna between the head and styloid process which help to retain the tendon of extensor carpi ulnaris in its track during pronation and supination. The retinaculum separates the extensor tendons from the superficial fascia, and by deep attachments forms a series of six tunnels for the extensor tendons. The most radial tunnel is that for the tendons of abductor pollicis longus and extensor pollicis brevis. The tendon of abductor pollicis longus is the most radial and is often in several strands. The extensor pollicis brevis tendon is relatively slender. There may be a septum separating it from the abductor. The ulnar partition of this tunnel is attached to an obvious ridge separating it from the next ulnarward tunnel for the tendons of extensor carpi radialis longus and brevis. The tendon of extensor carpi radialis brevis lies nearest to the ulnar partition of their tunnel, which is attached to the dorsal (Lister's) tubercle. The distal border of the retinaculum at this site is quite strong, giving it a firm edge. The tunnel for the tendon of extensor pollicis longus lies immediately to the ulnar side of the tubercle and runs a little obliquely to allow the tendon to change direction towards the radial side. Only a short partition separates the tunnel for extensor pollicis longus from the wider tunnel to its ulnar side for the tendons of extensor digitorum communis and extensor indicis, together with the termination of the anterior interosseous artery and the posterior interosseous nerve. The ulnar partition of this tunnel is attached to the posterior tubercle of the ulnar notch of the radius. The tendon of extensor digiti minimi lies in a tunnel over the distal radio-ulnar joint. The tendon of extensor carpi ulnaris lies in a tunnel over the dorsal surface of the head of the ulna, in a groove between it and the styloid process. It is retained by proximal deep fibres of the retinaculum forming the ulnar partition of the tunnel.

213. How do the proximal ends of the radius and ulna develop ?

i. At birth the head and neck of the **radius** are cartilaginous. The leading edge of the metaphysis as it invades the neck slopes a little distally and laterally so that on an anteroposterior X-ray film of the elbow the impression can be given of angulation of the bone. The secondary centre for the head appears at 5 years of age as a small flat and occasionally bipartite nucleus. As the head grows the epiphysis widens but remains flat. Fusion to the metaphysis occurs at 14 years of age.

ii. The coronoid process of the **ulna** ossifies with the shaft. It never develops a secondary centre. At birth the ulna is cartilaginous proximal to the mid-point of the trochlear notch. Usually two round or oval secondary centres appear in the olecranon at 9 years of age. The centres lie one in front of the other. The posterior centre is the larger, enveloped by the insertion of the triceps. The anterior centre underlies the upper quarter of the articular surface. Fusion of the epiphyseal centres to the metaphysis occurs at the age of 14. Sometimes the posterior part of the epiphyseal line may persist into adult life, suggesting an undisplaced fracture. Occasionally a centre develops in the insertion of the triceps tendon and persists into adult life. It can articulate with the trochlea and is known as a **patella cubiti**.

The proximal epiphyses of the radius and ulna contribute 20–30% of the growth of the forearm bones, and about 10% of the length of the limb.

214. How do the lower ends of the radius and ulna develop ?

i. The epiphyseal centre for the lower end of the **radius** appears at the age of about 1 year. A separate centre can occur in the tip of the styloid process, and if it persists the appearance may be mistaken for a fracture. Fusion to the metaphysis occurs between 16 and 18 years.

ii. The epiphyseal centre for the lower end of the **ulna** appears at the age of between 6 and 7 years. There is no separate centre for the styloid process. Fusion to the metaphysis occurs at about the same time as for the radius.

The epiphyseal centres for the lower ends of the radius and ulna contribute 75–80% of the growth of the forearm bones, and about 40% of the growth of the limb.

WRIST AND HAND

When applied to the wrist and hand, the terms anterior, posterior, medial and lateral, usually become palmar, dorsal, ulnar and radial respectively, and are so used in the following answers. Some surgeons prefer the term volar to palmar, for instance when describing the cartilaginous plates of the metacarpophalangeal and interphalangeal joints.

215. What bony parts can be felt about the wrist ?

They are —
  the tubercle of the scaphoid.
  the palmar crest of the trapezium.
  the tip of the styloid process of the radius.
  the waist of the scaphoid distal to the styloid process of the radius in the floor of the anatomical snuff box.
  the proximal end of the first metacarpal at the apex of the snuff box.
  the dorsal (Lister's) tubercle of the radius.
  the adjoining dorsal prominences of the bases of the second and third metacarpals.
  the dorsal pole of the lunate.
  the dorsal surface of the triquetral.
  the head and styloid process of the ulna.
  the pisiform.
  the tip of the hook of the hamate. Firm palpation of this bony point may cause discomfort by pressure on the superficial branch of the ulnar nerve, which is close to or overlying it.
  the transverse palmar ridge of the radius near the articular margin.

216. Where are the palmar wrist skin creases in relation to the underlying bones ?

The **proximal** skin crease overlies the wrist joint. A more accurate representation of the surface projection of the joint is a line gently concave downwards joining the radial and ulnar styloid processes.

The **distal** skin crease overlies the distal scaphoid, the head of the capitate, the body of the triquetral and the pisiform.

217. What structures comprise the articular surfaces of the wrist joint ?

The **proximal** surface is a transverse concave ellipsoid formed by the distal articular surface of the radius and the distal surface of the triangular fibrocartilaginous disc, the proximal surface of which is part of the inferior radio-ulnar joint. The radial surface is bisected by a low ridge into two concavities, a lateral one for the scaphoid and a medial one for the lunate. The **distal** surface is a transverse convex ellipsoid made by the smooth proximal articular surfaces of the scaphoid, lunate and triquetral and their interosseous ligaments.

218. What features of an X-ray film of the wrist joint can be used to assess the alignment of the proximal articular surface?

  i. On a lateral view the angle between the plane of the distal articular surface of the radius, drawn between the dorsal and palmar margins, and a line perpendicular to the axis of the shaft of the radius, is called the angle of palmar tilt. It averages about 10°.

ii. On a dorso-palmar view the angle between the plane of the distal articular surface of the radius, drawn between the tip of the styloid process and the distal margin of the ulnar notch, and a line perpendicular to the axis of the shaft of the radius, is called the angle of radial inclination. It averages about 22°.

iii. Also on a dorso-palmar view the distance between two parallel lines drawn perpendicular to the axis of the radius, one passing through the tip of the styloid process of the radius and the other tangentially to the distal outline of the head of the ulna, is called the radial length. It averages about 9 mm.

These measurements can be used to assess the degree of displacement of the distal fragment of the radius in the common Colles fracture.

219. What features of a dorso-palmar X-ray view of the carpus can be used to assess the proper alignment of its constituent bones ?

a) The outlines of the carpal articular surfaces form three arcs. The proximal or first arc is traced along the proximal convex surfaces of the scaphoid, lunate and triquetral. The second arc is traced along the distal concave surfaces of the same bones. The distal or third arc is traced along the proximal convex surfaces of the capitate and hamate. These arcs should be smooth and uninterrupted. A break in an arc implies ligamentous disruption.

b) The spaces between the bones are not marked. A wide gap, for instance between the scaphoid and lunate, indicates disruption of the scapholunate ligaments. However, an apparent gap between the scaphoid, capitate and trapezoid may indicate the site of a cartilaginous supernumary 'bone', an os centrale.

c) The shape of the bones is characteristic. If, for instance, the outline of the lunate appears triangular rather than trapezoid, it may be dislocated.

220. What features of a lateral X-ray view of the carpus can be used to assess the proper alignment of its constituent bones ?

i. The head of the capitate normally fits concentrically into the concavity of the lunate, and the proximal surface of the lunate fits into the concavity of the radial articular surface. If this alignment is not carefully checked a dislocation of the lunate can be missed.

ii. The angle between a line drawn perpendicularly to a line joining the poles of the lunate, and one along the axis of the scaphoid is called the scapholunate angle. It is on average about 50° (between about 30–60°). An angle greater than 70° suggests disruption of scapholunate ligaments.

221. What are the ligaments of the wrist joint ?

The **palmar ligaments** of the wrist joint are thick and strong. They are best seen from within the joint. They are —

a) the following four bands which arise from the radius, in order from lateral to medial —

 i. the so-called **radial collateral** ligament. It is not truly collateral, but arises fron the palmar surface of the radial styloid process near its articular margin and passes to the tubercle of the scaphoid and to the fibrous tunnel for the tendon of the flexor carpi radialis through the flexor retinaculum. It passes in front of the transverse axis of the joint. If it were truly collateral ulnar deviation of the wrist would not be possible.

 ii. the **radiocapitate** ligament. It arises from the palmar surface of the styloid process medial to the radial collateral ligament and passes obliquely medially to the capitate, with a few fibres attaching to the scaphoid on the way.

 iii. the **radiolunate** ligament. It is very strong, almost transverse, and passes from just medial to the radiocapitate ligament to the palmar pole of the lunate.

 iv. the **radioscapholunate** ligament. It arises from the articular margin of the radius where the sagittal ridge of its articular surface meets the margin, and inserts into adjacent fossae on the proximal and palmar facing surfaces of the scaphoid and lunate. It is deep to the radiolunate ligament.

b) on the ulnar side is the **ulnolunate** ligament. It arises from the palmar border of the triangular cartilage and passes to the palmar pole of the lunate.

The palmar ligaments of the wrist joint radiate in a distal V-shaped manner from radial and ulnar sides to the centre of the carpus, towards the dorso-palmar axis of the movements of radial and ulnar deviation (abduction and adduction). The ligaments pass predominantly from radial to ulnar sides preventing ulnar displacement of the carpus in the position of function of the hand.

The **dorsal ligaments** comprise three relatively weaker bands passing obliquely medially and distally from the dorsal rim of the radius to the scaphoid, lunate and triquetral. They are reinforced by the floors and septa of the six dorsal wrist compartments.

The **ulnar carpal meniscus**, when well developed, is a crescentic band of fibrous tissue on the ulnar side of the joint. Its concave free edge juts into the wrist joint cavity. Its convex base is continuous with the capsule. The dorsal horn of the meniscus arises from the dorsal corner of the ulnar notch of the radius with the attachment of the triangular cartilage. Its palmar horn attaches to the palmar surface of the triquetral just proximal to the articular surface for the pisiform. It covers the tip of the ulnar styloid, which is thus enclosed in a space, free of ligamentous attachment, called the **prestyloid recess**, lined by synovium, between the meniscus and the triangular cartilage.

There is no ligament attached to the styloid process of the ulna so that there is no true ulnar collateral ligament of the wrist joint. What is sometimes described as an ulnar collateral ligament is actually capsule thickened by fibres of the extensor retinaculum and the floor of the sheath of the extensor carpi ulnaris.

Dynamic collateral support is provided by extensor carpi ulnaris tendon on the ulnar side, and by abductor pollicis longus and extensor pollicis brevis tendons on the radial side.

222. What are the factors that contribute to the stability of the wrist joint ?

The concave articular surface of the radius is relatively shallow, offering little support against displacement of the carpus. The main support to the wrist is ligamentous. **Palmar displacement** is resisted by the strong palmar ligaments. **Dorsal displacement** is resisted by the dorsal ligaments which, though weaker, are strongly reinforced by the floors and septa of the compartments of the extensor retinaculum. **Radial displacement** is resisted to some degree by the overhang of the radial styloid. The palmar ulnolunate ligament also resists radial displacement. Although the dorsal and palmar radiocarpal ligaments pass predominantly in an ulnar direction, radial displacement of the carpus entails some distraction of the surfaces, and thus tightening of these ligaments. In addition the tendons of abductor pollicis longus and extensor pollicis brevis provide dynamic support to the radial side of the wrist. **Ulnar displacement** would appear to be the most vulnerable, there being no bony support. It is resisted strongly, however, by the ulnar direction of the palmar and dorsal radiocarpal ligaments, by the sling effect of the ulnar meniscus passing from the dorsal margin of the radius to the triquetral, and by the capsular thickening often described as the ulnar collateral ligament. The extensor carpi ulnaris tendon provides some dynamic support to the ulnar side of the wrist.

223. What movements occur at the wrist ?

Movements at the wrist are described in four directions —
Palmar flexion or flexion from neutral (0°) to 80°.
Dorsiflexion or extension from neutral (0°) to 70°.
Radial deviation (or abduction) from neutral (0°) to 20°.
Ulnar deviation (or adduction) from neutral (0°) to 30° in pronation, and a little more in supination.

224. What movements occur within the carpus ?

Traditionally the carpus has been regarded as simply forming a transverse joint between its proximal and distal rows of bones, accessory to the wrist joint, to increase the range of palmar and dorsiflexion of the hand on the forearm. This concept is now being revised in the light of an increasing realisation of the considerable complexity of the interaction between the multiple joint surfaces and ligaments that join the radius and ulna to the metacarpals.

In summary the carpus is now thought to function as three longitudinal columns —

i. The **central column** is regarded as being formed by the lunate, capitate and the other bones of the distal row. The joints between the radius and lunate, and lunate and capitate are relatively lax. The joints between the bones of the distal row and between them and the second to the fifth metacarpals are relatively rigid. This column of bones provides the central flexion–extension link between forearm and hand.

ii. The **lateral column** comprises the scaphoid. It forms a stabilising link across the lunocapitate joint which would otherwise buckle under compression. It can rotate on a transverse axis to adjust the structure of the carpus to allow radial deviation of the hand. In ulnar deviation and in extension, the axis of the scaphoid is directed longitudinally, and in this position lunocapitate or mid-carpal movement is eliminated. In radial deviation and in flexion, the long axis of the scaphoid becomes perpendicular to the long axis of the radius as the bone flexes into a foreshortened attitude. In this movement the lunocapitate joint unlocks and allows a greater range of flexion than at the wrist joint.

iii. The **medial column** consists of the triquetral and pisiform. The pisiform has a dynamic stabilising function by way of its tendinous, muscular, ligamentous and retinacular attachments. The triquetral allows controlled shortening of the ulnar side of the carpus in ulnar deviation by distal telescopic migration on the hamate. It is also thought to be the centre of some slight independent pronation and supination movements of the carpus on the radius.

225. What is the blood supply of the scaphoid ?

Most of the surface of the scaphoid is covered by articular cartilage, in contact with the radius, lunate, capitate, trapezoid and trapezium. The non-articular surface for the attachment of ligament, retinaculum and muscle, and for the entry of blood vessels, is confined to narrow palmar and dorsal strips which link distally with the indented lateral surface or waist of the bone, and the distal palmar surface and tubercle. The proximal pole of the bone is almost entirely intra-articular, like the head of the femur. Although the scaphoid limb of the radioscapholunate ligament attaches to the proximal pole it carries no significant blood supply to it. The scaphoid is supplied from branches of the radial artery as it spirals round the lateral side of the wrist, in the floor of the anatomical snuffbox, to the first interosseous space. Branches to the bone enter the ligamentous areas in three groups. A **proximal group** enters the dorsal ligamentous area, though often these branches do not extend to the proximal end of the strip. In one in seven subjects no branches pass to the dorsal strip. A **middle group** of vessels enters the waist of the bone just distal and lateral to the radial articular surface. These vessels provide the main supply to the proximal two thirds of the bone. A **distal group** enters the distal part of the palmar aspect of the bone at the base of the tubercle and supplies only a localised area around it. A fracture of the proximal third of the scaphoid is liable to cause avascular necrosis of the proximal fragment in up to a third of cases.

226. What are the features of the flexor retinaculum ?

The flexor retinaculum forms the roof of the carpal tunnel. It is a transverse band attached laterally to the tubercle of the scaphoid and the palmar crest of the trapezium, and medially to the pisiform and the hook of the hamate. The distal border of the retinaculum is thick and well defined where fat surrounds the superficial palmar arterial arch. The proximal border is not well defined, being continuous with the deep fascia of the forearm. A proximal, medial, superficial part of the retinaculum, also called the **volar carpal ligament,** is a thickening of the deep fascia arising from its fusion with the distal part of the tendon of flexor carpi ulnaris and the pisiform, and merging laterally with the main part of the retinaculum. Its distal border forms a hiatus for the emergence of the ulnar nerve and vessels from under the deep fascia. The fibres of the retinaculum run in diverging directions from their origins, weaving across each other to form a dense feltwork. Many fibres pass transversely to merge with opposite fibres. Others cross obliquely and distally to become the fasciculi of origin of the thenar and hypothenar muscles. Some superficial vertical fibres run into the origin of the palmar aponeurosis. Its deep shiny aspect passes smoothly into the covering of the side walls and dorsal aspect of the carpal tunnel. The tendon of flexor carpi radialis lies at first superficial to its radial side, just medial to the tubercle of the scaphoid. As it descends it passes between the superficial layers of the retinaculum to become sandwiched between the fibrous radial wall of the carpal tunnel and the scaphoid proximally, and the groove between the trapezium and trapezoid distally. Passing down superficial to the retinaculum and medial to the tendon of flexor carpi radialis, in order from lateral to medial sides, are the palmar branch of the median nerve, the tendon of palmaris longus and the root of the palmar aponeurosis, the palmar branch of the ulnar nerve, and the ulnar vessels and ulnar nerve in the space of Guyon. The median nerve is in direct contact with the deep aspect of the retinaculum. It usually divides at the distal border of the retinaculum into two or more branches, and the important motor branch of the nerve to the thenar muscles usually arises from the lateral side of the parent nerve, to recurve round the distal border of the retinaculum. The muscular branch may arise more proximally under the retinaculum and run in a tunnel through it to emerge 2–6 mm proximal to the distal border. It may arise from the ulnar side of the parent nerve and cross in front of it in contact with the deep surface of the retinaculum. These variations need to be kept in mind during the operation to decompress the carpal tunnel when the retinaculum is incised longitudinally.

227. What is the surface marking of the motor branch of the median nerve to the thenar muscles ?

The usual point of emergence of the motor branch from behind the retinaculum is at the intersection of the thenar flexion crease and **Kaplan's cardinal line**. This line passes from the middle of the interdigital fold between the thumb and index finger towards the ulnar side of the hand, parallel with the proximal palmar crease, and usually passing about 4–5 mm distal to the pisiform.

228. What are the features of the carpal tunnel ?

The carpal tunnel is the unyielding space enclosed by the arch of the carpal bones and the flexor retinaculum. Through it pass the nine tendons of flexor digitorum longus, flexor digitorum superficialis and flexor pollicis longus, and the median nerve. The tendon of flexor carpi radialis passes the carpus in the distal radial wall of the tunnel and is not contained within it. The pressure within the tunnel in the neutral position is about 3–4 mm Hg. The upper limit of normal is taken to be 5 mm Hg. In palmar flexion it rises to about 8–10 mm Hg. In dorsiflexion it rises to about 11–13 mm Hg. It has been found experimentally that an external pressure of 20–30 mm Hg causes a retardation of venular flow in the epineurium, and that a pressure of about 80 mm Hg causes a complete arrest of intraneural blood flow. Patients who present with carpal tunnel syndrome, that is with impaired median nerve conduction in the carpal tunnel, have a raised resting pressure within the tunnel of around 10–30 mm Hg. A clinical test for carpal tunnel syndrome (Phalen's test) is to flex the wrist for a minute, when the typical paraesthesiae of the syndrome are induced in the median nerve distribution. In these patients the pressure within the carpal tunnel is found to reach 90–100 mm Hg. The paraesthesiae also typically occur at night, when in the same patients the pressure within the tunnel can reach 40–50 mm Hg. This is due probably to the redistribution of fluid in the horizontal position, cessation of the forearm muscle-pump action, and perhaps a tendency to sleep with the wrist flexed. The distal border of the retinaculum is the thicker and it is here that maximum delay in conduction of the nerve seems to be incurred in carpal tunnel syndrome. Fractures or dislocations, or compressive injuries at or near the carpal tunnel can often precipitate the onset of a compression syndrome.

229. What is Guyon's canal ?

Guyon's name is attached to an area of the proximal part of the hypothenar eminence which has become recognised as a site of numerous pathologies causing entrapment of the ulnar nerve. The nomenclature of the region is not yet standardised. It is perhaps better referred to as a space. The space of Guyon can then be regarded as roofed, on its palmar aspect, from superficial to deep, by skin reinforced by a dense subcorial layer of fibrous tissue, a pad of fat, and the palmaris brevis. Deeply, forming its floor, from lateral to medial, is the medial part of the flexor retinaculum, the pisohamate and pisometacarpal ligaments, and the fascia over abductor digiti minimi. The ulnar neurovascular bundle enters the space from above through a hiatus in the deep fascia immediately lateral to the pisiform. The superficial rim of the hiatus is the distal border of the volar carpal ligament arising from the distal part of the tendon of flexor carpi ulnaris and the pisiform. This fuses laterally with the deep rim of the hiatus which is the proximal border of the flexor retinaculum, also arising from the pisiform. The medial limit to the space is the line of merger of the fibres of palmaris brevis and the fascia over abductor digiti minimi. The lateral limit is at the site of origin of proximal fibres of palmaris brevis from the flexor retinaculum where the palmaris longus tendon begins to spread out into the palmar aponeurosis. Distally the space merges with the subcutaneous tissue of the distal part of the hypothenar eminence. As the ulnar nerve and vessels pass through the hiatus to enter and descend the space they become embedded in soft fat. The nerve lies medial to the vessels. In the space the ulnar nerve and vessels divide into their deep and superficial branches. A tunnel for the deep neurovascular branches passes dorsally from the floor of the space, as defined above, under a fibrous arcade between the pisiform and the tip of the hook of the hamate which gives rise to fibres of abductor digiti minimi. The tunnel passes over the pisometacarpal ligament distally, and then deeply and laterally around the distal margin of the hook of the hamate, between the two heads of origin of opponens digiti minimi, to enter the deep part of the palm. This tunnel, rather than the space as defined above, is better referred to as Guyon's canal. A ganglion developing in relation to the pisotriquetral joint could invade the space of Guyon and compress the deep branch of the ulnar nerve.

230. How are the tendons inserted that primarily flex the wrist ?

The tendon of **flexor carpi radialis** is inserted into the palmar surfaces of the adjoining bases of the second and third metacarpals. The tendon descends at first superficial to the lateral side of the flexor retinaculum, just medial to the tubercle of the scaphoid. It then sinks into the lateral wall of the carpal tunnel, to pass to the medial side of the palmar crest of the trapezium, and then in line with the joint between the trapezium and trapezoid, to its insertion beyond the floor of the carpal tunnel.

The tendon of **flexor carpi ulnaris** is inserted distal to its sesamoid, the pisiform, by two bands, the pisohamate ligament into the base of the hook of the hamate, and the pisometacarpal ligament mainly into the palmar and ulnar aspects of the base of the fifth metacarpal, and to a lesser extent into the palmar surface of the base of the fourth. Some fibres of flexor carpi ulnaris pass laterally into the superficial part of the retinaculum or volar carpal ligament and some seem to pass distally into the fibres of origin of abductor digiti minimi.

231. Where are the tendons inserted that primarily extend the wrist ?

The tendon of **extensor carpi radialis longus** is inserted into a rough facet over the radial side of the dorsal surface of the base of the second metacarpal. Sometimes a substantial radial slip of the tendon diverges to attach to the ulnar side of the base of the first metacarpal, and may incorporate part of the intermetacarpal ligament.

The tendon of **extensor carpi radialis brevis** is inserted into the dorsal aspect of the styloid process of the base of the third metacarpal.

The **tendon of the extensor carpi ulnaris** is inserted into a tubercle on the dorso-ulnar aspect of the base of the fifth metacarpal.

232. What is the anatomical snuff box ?

The anatomical snuff box is the surface depression in the skin on the radial side of the wrist, just distal to the styloid process of the radius, bounded by the tendons of extensor pollicis brevis and extensor pollicis longus. Deep to its floor lie successively the radial styloid, the scaphoid, and the trapezium. Running distally and dorsally across the floor is the radial artery giving off branches to the carpus. Running across the snuff box superficial to the tendons, are one or two branches of the superficial radial nerve. They can be felt passing over the tendon of extensor pollicis longus.

233. What are the movements of the thumb ?

The classic movements described are —

i. **flexion**, in which the thumb is carried across the palm in the plane of the metacarpals.

ii. **extension**, the opposite movement.

iii. **abduction**, in which the thumb is carried away from the palm at right angles to the plane of the metacarpals.

iv. **adduction**, the opposite movement.

v. **opposition** in which the thumb is flexed and abducted, and at the same time rotated, so that the thumb nail is in the same plane as the nail of a finger.

vi. **reposition**, the opposite movement.

vii. **circumduction**, a composite of the above movements in turn, to describe an arc of a cone.

Abduction and adduction are also sometimes described as movements of the thumb away from or towards the second metacarpal, from whatever position the thumb may be in. One can then describe abduction in the above sense as palmar abduction, and extension as radial abduction.

234. Where are the tendons of the extrinsic muscles of the thumb attached ?

The tendon of the **abductor pollicis longus** often consists of several fasciculi. The main insertion of the tendon is into a small tubercle on the radial side of the base of the first metacarpal. Other slips may pass into the origin of abductor pollicis brevis and opponens pollicis, and the radial or dorsal aspect of the trapezium. It has some flexor as well as abductor action on the metacarpal at the trapezio-metacarpal joint.

The tendon of **extensor pollicis brevis** is a slender one, and is inserted into the dorsal aspect of the base of the proximal phalanx. An extension of the tendon may insert into the base of the distal phalanx. It abducts the thumb a little as well as extending it.

The tendon of **extensor pollicis longus** is inserted into the dorsal aspect of the base of the distal phalanx. It receives expansions from abductor pollicis brevis and adductor pollicis. While a prime mover in extension of the thumb it assists in abduction and reposition.

The tendon of **flexor pollicis longus** inserts into the proximal palmar aspect of the base of the distal phalanx.

235. What are the features of the trapezio-metacarpal joint ?

The trapezio-metacarpal joint is essential for the proper mobility of the thumb, and therefore of the function of the whole hand.

i. It is a very mobile joint. Though strong the capsule is loose and easily distensible. The articular surfaces are relatively incongruent, being separate at their periphery in the neutral position.

ii. The distal articular surface of the trapezium faces distally, radially (20°) and palmarwards (35°). The articular surfaces are saddle-shaped. Their geometry is such that the optimum plane of flexion and extension of the thumb, which is in the direction of the greatest convexity of the trapezium and of the concavity of the metacarpal base, is inclined some 15° to the transverse plane of the hand in a palmar and ulnar direction. Conversely the optimum plane of abduction and adduction of the thumb, which is in the direction of the greatest concavity of the trapezium and of the convexity of the metacarpal base, is inclined some 15° to the sagittal plane of the hand in a palmar and radial direction.

iii. Flexion and abduction of the thumb are accompanied by passive (conjunct) opposition of the metacarpal at the trapezio-metacarpal joint. Likewise extension and adduction are accompanied by conjunct reposition. This is brought about because ligamentous thickenings of the capsule restrain the dorso-ulnar aspect of the base of the metacarpal more than the palmar-radial aspect, allowing the base of the metacarpal to swing round its more restrained corner. Thus a palmar oblique ligament runs from the distal end of the palmar crest of the trapezium to the palmar beak and ulnar side of the base of the first metacarpal; a dorsal oblique ligament runs from the ulnar dorsal tubercle of the trapezium to the ulnar tubercle of the base of the metacarpal; and palmar and dorsal intermetacarpal ligaments pass to the ulnar tubercle of the first metacarpal from the adjoining base of the second. It is also noticeable that the palmar-radial sectors of the opposing articular surfaces are somewhat flatter than those of the dorso-ulnar sectors, which allows more translation movement in the peripheral part of the radius of rotation.

iv. The close-packed position of the joint, of maximum stability, in flexion, abduction and maximum opposition, occurs in power pinch and grip. The metacarpal is fixed by the pull of the thenar muscles and abductor pollicis longus, and by the dorsal oblique and intermetacarpal ligaments. The area of maximal stress through the joint appears to be in the radiopalmar quadrant, where the degenerative changes of osteoarthritis are seen at their worst.

236. What do the thenar muscles do ?

The thenar muscles can be considered to be arranged in two groups. A radial group consists of the abductor pollicis brevis, opponens pollicis and flexor pollicis brevis. An ulnar group consists of the two heads of the adductor pollicis.

Abductor pollicis brevis arises mainly from the palmar surface of the flexor retinaculum, the tubercle of the scaphoid and the palmar crest of the trapezium, but it also receives fibres from the skin of the thenar eminence and the tendons of abductor pollicis longus and palmaris longus. It inserts into the radial sesamoid in the palmar plate and the side of the capsule of the metacarpophalangeal joint of the thumb, into the radial tubercle of the proximal phalanx of the thumb, and into the extensor hood over the joint. **It is active in all movements in which the thumb metacarpal moves away from the second, and in opposition.**

Opponens pollicis arises from the palmar aspect of the flexor retinaculum deep to the origin of abductor pollicis brevis, proximal to the fibres of origin of the superficial head of the flexor pollicis brevis. The only muscle inserted directly into the thumb metacarpal, it attaches to the lateral surface of the bone, covering in its course the palmar aspect of the trapeziometacarpal joint. **It is active during both abduction and opposition.**

Flexor pollicis brevis arises by two heads. The superficial head arises from the distal part of the palmar surface of the flexor retinaculum, distal to the fibres of origin of the opponens, and deep to the abductor, and from the distal edge of the retinaculum. The deep head arises from the adjacent palmar surfaces of the capitate and trapezoid in the distal part of the floor of the carpal tunnel, proximal to the origin of the oblique head of the adductor pollicis. It crosses deep to the tendon of flexor pollicis longus to join the superficial head. The two heads combine to insert into the radial sesamoid and capsule of the metacarpophalangeal joint and the radial side of the base of the proximal phalanx. **It appears to act primarily in opposition.**

Adductor pollicis also arises by two heads. An oblique head arises from the adjacent palmar surfaces of the capitate and trapezoid, distal to the origin of the deep head of the flexor pollicis brevis. The transverse head arises from the palmar surface of the base, and from the palmar crest as far as the neck, of the third metacarpal. The two heads combine to insert into the ulnar sesamoid and the ulnar side of the capsule of the metacarpophalangeal joint, the ulnar tubercle of the base of the proximal phalanx and the ulnar side of the extensor hood over the joint. **It is active in any movement approximating the first to the second metacarpal.**

237. What are the features of the metacarpophalangeal joints ?

Those of the **fingers** have the following features —

a) **Articular surfaces**. Viewed end-on the proximal articular surface is a slightly transversely flattened condyle, but towards the palmar surface it becomes somewhat bicondylar. The distal surface of the joint is a concave transverse ellipsoid. It covers the whole of the proximal surface in its transverse diameter, but only a third of its surface in its sagittal diameter. The surfaces allow flexion and extension, abduction and adduction and a little axial rotation.

b) A **dorsal capsule**, which is thin, almost bursa-like proximally, but strengthened distally by a deep expansion from the long extensor tendon.

c) A **palmar capsule** which is thickened to form a complex fibrous palmar plate. Its dorsal surface forms part of the articular surface. Its palmar surface forms the proximal part of the fibrous flexor tunnel. The A1 band of the sheath is attached to each margin of the plate (see question 244). The sides of the plate are also continuous with strong ligaments between adjacent plates, forming with them the deep transverse ligament of the palm stretching from the radial side of the index finger to the ulnar side of the little finger. There may be sesamoid bones within and to the sides of the palmar plates. The most common one is to the ulnar side of the plate of the little finger joint. The plate is firmly attached along the palmar surface of the proximal border of the proximal phalanx, but loosely to the neck of the metacarpal.

d) The **lateral capsule**, which is thickened on each side to form two important structures —

i. the proper collateral ligament. It is large and triangular. Arising by its apex from the dorso-lateral tubercle of the head of the metacarpal and the conspicuous fovea antero-distal to it, it radiates distally to insert into the prominent lateral condyle of the proximal phalanx. The radial ligament is stronger than the ulnar one.

ii. the thin accessory collateral ligament. It is attached proximally with the proper collateral ligament. It fans out to attach to the lateral margin of the palmar plate.

The collateral ligaments of the metacarpophalangeal joints are tightest in flexion because of the slight cam effect of the contour of the proximal articular surface. Metacarpophalangeal joints should therefore be splinted in flexion rather than extension, since contraction of the ligaments in extension would cause difficulty in flexion on mobilisation. The opposite is true for the interphalangeal joints.

The metacarpophalangeal joint of **the thumb** has a bicondylar configuration, similar to an interphalangeal joint. Sesamoid bones in the palmar plate cause slight impressions on the condyles of the metacarpal head. The ulnar sesamoid receives the fibres of insertion of adductor pollicis. The radial sesamoid receives the insertion of flexor pollicis brevis and partly of abductor pollicis brevis.

238. What are the features of the interphalangeal joints ?

a) The **articular surfaces**. Viewed end-on the proximal surface of an interphalangeal joint presents a bicondylar configuration within a trapezoid outline, the palmar margin being wider than the dorsal. The articular surface is more extensive on the palmar aspect than the dorsal, signifying a greater range of flexion. The distal surface of the joint shows reciprocal condylar depressions separated by a median intercondylar ridge. The surfaces lack complete congruity, thereby allowing some translation and rotational movements necessary to adapt the digits to irregular shapes in power grip. The surfaces are set so that in flexion the tips of the fingers point in common to the base of the thumb.

b) The **dorsal capsule**. There is no formal capsule intervening between the synovial membrane and the extensor apparatus dorsal to the collateral ligaments. A pouch of synovium spreads a little proximally deep to the tendon blending with periosteum and areolar tissue. Division of the extensor tendon over the joint immediately involves the synovial cavity.

c) The **lateral capsule** is thickened to form —

i. the proper collateral ligament, which is attached proximally to a concavity towards the dorsal aspect of the lateral face of the condyle. It fans out distally to attach around the palmar lateral tubercle of the base of the phalanx distal to it, spreading distally beyond the area of the epiphyseal plate into the metaphysis.

ii. the thin accessory collateral ligament, which is attached proximally with the proper collateral ligament. It also fans out to attach to the lateral margin of the palmar plate. It is tight in full extension but balloons out a little in flexion.

d) The **palmar capsule** is thickened to form the palmar plate. Its dorsal surface forms part of the articular surface. The palmar surface is covered by parietal tenosynovium of the flexor tendon sheath. It arises proximally by two lateral extensions in a swallow's tail configuration. Each tail is attached to bone along the lateral edge of the palmar surface of the phalanx proximal to it, within the distal part of an anular pulley of the flexor sheath. The proximal crescentic edge of the plate formed by the union of the tails of origin is separated from bone by a small synovial pouch. Distally the lateral margins of the plate receive the fibres of the accessory collateral ligaments and insert strongly into the palmar tubercles of the phalanx distal to it. The central part of the plate has a weaker insertion into the palmar surface of the base of the distal bone. The dorsal surface of the distal part of the plate presents a low transverse meniscal edge projecting into the joint.

The collateral ligaments and the palmar plate of an interphalangeal joint are under greatest tension in extension, and resist hyperextension. Interphalangeal joints should therefore preferably be splinted in extension rather than in flexion, since contraction of the ligaments in flexion would cause difficulty in extension on mobilisation. This position contrasts with that of the metacarpophalangeal joints.

239. How are the flexor tendons inserted into the phalanges of the fingers ?

Each tendon of the **flexor digitorum profundus** inserts into the proximal palmar aspect of the base of a terminal phalanx. As the tendon passes through the split in the tendon of superficialis its fibres undergo a bilateral spiralisation. Seen from the palmar aspect the superficial fibres diverge from the midline to spiral distally round the sides of the tendon to the midline of the dorsal aspect, where they enter the central part again before continuing their spiral course to spread out into their insertion. The distal part of the tendon appears grooved lengthways in the midline.

Each tendon of the **flexor digitorum superficialis** inserts into the middle phalanx in a somewhat complicated fashion. Within the proximal part of the fibrous sheath its fibres split into two bands which spiral round to the dorsal aspect of the profundus tendon, and there partially cross over to form the chiasma of the tendon. The distal bands of the chiasma taper to insert along the lengths of the lateral borders of the palmar surface of the middle phalanx deep to the attachment of the A4 pulley of the fibrous sheath, nearly to the neck of the phalanx (see question 244).

240. What are the lumbrical muscles and what do they do ?

The four lumbricals arise in the palm from the tendons of the flexor digitorum profundus. The radial two lumbricals are fusiform and arise from the radial and palmar surfaces of the tendons to the index and middle fingers; the ulnar two are pennate and arise from the adjacent borders and palmar surfaces of the tendons of the middle, ring and little fingers. Each passes distally along the radial side of a profundus tendon, palmar to the deep transverse ligament, and palmar to the axis of flexion of the metacarpophalangeal joint, at which level a narrow flat tendon is formed which joins the radial border of the extensor expansion. They vary in size between fingers, between hands and between individuals. They may be absent, or they may be strong, measuring up to 8 cm long and 8 mm in diameter. They may insert on the ulnar side of the extensor apparatus or arise at a more proximal level.

A lumbrical acts —

i. to flex the metacarpophalangeal joint only in resisting hyperextension by extensor digitorum, or when the interphalangeal joints are extended.

ii. to extend the interphalangeal joints, contracting with extensor digitorum.

iii. to slacken the distal part of the tendon of profundus, improving the efficiency of extension of the interphalangeal joints by extensor digitorum.

iv. probably as a fine co-ordinator of muscle action by virtue of the impressive density of its proprioceptive nerve endings. The variability of the muscle, however, seems at odds with the conception of this as the very important function often proposed.

Lumbricals do **not** contract in simple flexion of metacarpophalangeal joints, and seem not to assist in radial deviation of the fingers.

241. How are the long extensor tendons inserted into the phalanges ?

The long extensor tendons of the fingers (comprising those of extensor digitorum communis, extensor indicis and extensor digiti minimi) join with the insertions of the intrinsic muscles to form complex tendinous plexuses or expansions, which are held in position by equally complex systems of thin ligaments and fascia. On the dorsum of each finger the terminal part of the extrinsic tendon forms the central part of the extensor apparatus. Over the dorsum of the metacarpophalangeal joint a deep central slip or layer of the tendon fuses with the capsule of the joint and thereby gains attachment to the base of the proximal phalanx. It is tight only on hyperextension of the joint. Over the shaft of the proximal phalanx the tendon divides into a middle band that inserts into the dorsal aspect of the base of the middle phalanx and two lateral bands which continue distally and meet to insert similarly into the base of the distal phalanx.

242. What are the interosseous muscles and what do they do ?

There are two groups of interossei. There are four pennate dorsal interossei which arise from the sides of the shafts of adjacent metacarpals. They are numbered from the radial side. There are usually described only three unipennate palmar interossei which arise from the palmar borders of the shafts of the metacarpals — from the ulnar side of the second, from the radial side of the third and from the radial side of the fourth. They also are numbered from the radial side. A variably present small muscular belly which arises from the palmar surface of the base of the first metacarpal, and inserts with flexor pollicis brevis, is said by some to constitute a first palmar interosseous. The tendons arising from these muscles to the fingers pass distally dorsal to the deep transverse ligament of the palm, which separates them from the tendons of the lumbricals.

An interosseous tendon inserts into —

i. the capsule of a metacarpophalangeal joint at the distal border of the palmar plate.

ii. the lateral tubercle of a proximal phalanx.

iii. the interosseous hood of the extensor apparatus. This is formed by a fanning out of the tendon. Proximally the fibres run transversely to merge with the extrinsic tendon. More distally the fibres run obliquely distally to merge with the side of the extrinsic tendon, and with the central band and so into the base of the middle phalanx. The most distal fibres merge with the lateral bands of the extrinsic tendon. The lateral band is joined on the radial side distally by the tendon of a lumbrical.

The emphasis of insertion differs between the groups of interossei —

i. The three ulnar **palmar** interossei insert into the ulnar side of the index and the radial sides of the ring and little fingers so that they **adduct** the fingers to the middle finger.

ii. The **dorsal** interossei insert into the radial side of the index and middle fingers and the ulnar sides of the middle and ring fingers so that they **abduct** the fingers from the line of the middle finger.

iii. The first dorsal interosseous inserts predominantly into the radial side of the base of the proximal phalanx of the index finger, with only a small contribution to the proximal part of the interosseous hood. It provides mainly strong abduction against an opposed thumb.

iv. The other palmar interossei and the third dorsal interosseous insert predominantly distally into the lateral bands.

v. The second and fourth dorsal interossei insert evenly into the bases of the phalanges and the interosseous hoods.

The interossei —

i. abduct or adduct the fingers by their insertions into the proximal phalanges.

ii. flex the metacarpophalangeal joints by their insertions into the interosseous hoods.

iii. extend the proximal and distal interphalangeal joints by their insertions into the central tendon and lateral bands of the extrinsic tendons.

243. How is the extensor mechanism stabilised over the dorsum of a finger ?

At the level of the metacarpophalangeal joint, what is called a sagittal band extends volarly, deep to the interosseous hood, from each side of the extensor tendon to the deep transverse ligament of the palm, at its junction with the palmar plate.

At the level of the proximal interphalangeal joint there are several stabilising structures —

i. a dorsal retinacular ligament joins the lateral bands, as they approach each other, filling the triangular interval between them.

ii. lateral or transverse retinacular ligaments join the lateral bands of the extensor tendon to the fibrous sheath at its C1 level more proximally (see question 244).

iii. longitudinal cords (of Landsmeer) are variably developed bands of fibres extending obliquely distally and dorsally from the lateral ridges of the proximal phalanx deep to the lateral bands of the extensor tendon. They pass palmar to the axis of flexion of the joint, to join the lateral bands of the extensor tendon at the level of the shaft of the middle phalanx. They pass deep to the lateral retinacular ligaments. They link flexion of the distal interphalangeal joint with flexion of the proximal joint, and help prevent hyperextension of the proximal joint.

244. How are the fibrous flexor tendon sheaths constructed ?

The fibrous flexor sheath **of a finger** extends from the level of the head of the metacarpal to the base of the terminal phalanx. It forms a tunnel with the palmar surfaces of the phalanges and the metacarpophalangeal and interphalangeal joints, and is made up of arching fibres between the lateral borders of the phalanges and the palmar plates of the joints. The sheath varies in thickness, as does the direction and attachment of its fibres. Some fibres pass predominantly transversely to form anular pulleys. Some fibres pass obliquely across each other to form cruciate pulleys. A sequence of pulleys of the sheath is recognised, from proximal to distal —

a **thick A1** (first anular) pulley, attached to the palmar plate of the metacarpophalangeal joint and to the proximal part of the proximal phalanx. It is sometimes known as the vaginal ligament.

a **thick A2** (second anular) pulley, attached to the margins of the proximal half of the shaft of the proximal phalanx, but separated from the A1 pulley by a distinct interval. It has a prominent thick proximal opening but is thickest at its distal end.

a **thin C1** (first cruciate) pulley, a narrow strip at the distal end of the A2 pulley.

a **thin A3** (third anular) pulley, attached to the edges of the palmar plate of the proximal interphalangeal joint.

a **thin C2** (second cruciate) pulley, a narrow strip present in the interval between the A3 and A4 pulleys.

a **thick A4** pulley, attached to the middle of the middle phalanx. It is thickest in its middle.

a **thin C3** (third cruciate) pulley, at the distal end of the A4 pulley. It may only be an oblique band.

a **thin A5** pulley, attached to the margins of the palmar plate of the distal interphalangeal joint.

The pulleys are linked by thin parts of the sheath without demonstrable fibre patterns. These interpulley parts of the sheath superficially overlap the free edges of the thick pulleys (principally A1 and A2) before attaching to them, thereby leaving free edges of pulley within the sheath. Small recesses of the sheath are thus formed between the thin parts of the sheath superficially and the free edges of the pulley deeply. The most obvious recess occurs at the distal end of an A2 pulley.

The fibrous flexor sheath **of the thumb** consists of three pulleys, the sequence of which, from proximal to distal is —

the **A1** (first anular) pulley which is quite thick and arises from the palmar plate of the metacarpophalangeal joint and the proximal part of the proximal phalanx.

the **oblique** pulley which arises from the ulnar side of the base of the proximal phalanx and crosses to the neck of the proximal phalanx just short of the interphalangeal joint. Fibres of the tendon of adductor pollicis insert into it.

the **A2** (second anular) pulley which is quite thin and arises from the distal part of the proximal phalanx and the volar plate of the interphalangeal joint.

245. How are the flexor tendons nourished within the tendon sheath of a finger ?

Within the flexor sheath of a finger the two flexor tendons receive a blood supply from four anastomotic arches between the two palmar digital arteries. Each arch lies in the fibrous tissue between the dorsal part of the fibrous sheath and the periosteum of a phalanx. From each arch arteries penetrate the sheath and pass in specialised strands of mesotenon, called vincula, to the tendon. The vincula that arise from the arches are now designated V1 to V4 in sequence from proximal to distal. They vary in shape. A vinculum may take the form of a band or mesentery near the insertion of the tendon and is described as short, or it may be a tubular structure inserted more proximally and described as long. There is some variation in the incidence of occurrence and type at the different levels. The proximal arterial arch is situated at the level of the proximal third of the proximal phalanx. A strong V1 short vinculum passes to each lateral slip of the superficialis tendon as it nears the tendinous chiasma. A weak V1 long vinculum reaches the same slip more proximally. The second arterial arch lies just proximal to the proximal interphalangeal joint. There is a strong V2 short vinculum to each superficialis slip at the chiasma, and a weaker V2 long vinculum passes between the superficialis slips to the profundus tendon. The third arterial arch lies midway between the proximal and distal interphalangeal joints. Delicate V3 long vincula pass to the profundus tendon. The distal arterial arch lies just proximal to the distal interphalangeal joint. A V4 short vinculum passes to the distal end of the profundus tendon. It is often strong enough to maintain some continuity between the ends of the tendon if it is lacerated at this site, so that weak flexion of the distal interphalangeal joint is still possible.

THE PELVIS

246. What is the pelvic brim ?

The pelvic brim is the boundary of the pelvic inlet (see question 249), which separates the greater from the lesser pelvis. The greater, or false pelvis, is that above the pelvic brim, flanked by the wings of the ilia. The lesser or true pelvis is that below the pelvic brim, continuous above with the greater pelvis, but limited below by the pelvic floor. Posteriorly the brim is formed by the promontory of the sacrum. From behind forwards, on each side, the brim is made up of the lower border of the ala of the sacrum, the sacroiliac joint, the arcuate line round to the iliopubic eminence, the pecten pubis (pectineal line), the pubic tubercle and pubic crest. Anteriorly is the upper border of the symphysis pubis. The part of the brim formed by a hip bone constitutes a linea terminalis.

247. What structures cross the pelvic brim ?

The urachus, raising the median umbilical fold of peritoneum, ascends behind the pubic symphysis. Structures which cross the pelvic brim, on each side, in order from front to back, are the —

Anastomosing pubic branches of obturator and inferior epigastric vessels, which, if enlarged, constitute abnormal obturator vessels from the inferior epigastric vessels.

Obliterated umbilical artery, raising the medial umbilical fold of peritoneum.

Ductus deferens or round ligament with their vessels.

Ovarian vessels and their accompanying nerves and supportive connective tissue, together comprising the infundibulo-pelvic or suspensory ligament of the ovary.

Ureter.

Internal iliac vessels.

Third and fourth lumbar splanchnic nerves.

Obturator nerve.

Iliolumbar vessels.

Lumbosacral trunk.

Sympathetic trunk.

On the right side the appendix commonly lies across the pelvic brim. On the left side the descending colon becomes the sigmoid colon as it crosses the brim. The superior rectal vessels, which are the continuation of the inferior mesenteric vessels, descend across the left common iliac vessels, into the right limb of the sigmoid mesocolon. Also on the left side parasympathetic nerves ascend over the brim to the descending colon.

The superior hypogastric plexus and the median sacral vessels descend across the sacral promontory.

248. How, briefly, is the pelvis constructed to withstand the forces applied to it ?

The pelvis is strongly constructed to —

i. withstand the weight of the body above it, which tends to thrust the upper part of the sacrum downwards between the hip bones, and thus to rotate the sacrum forwards.

ii. transmit the weight of the body above it to the ischial tuberosities and lower limbs.

iii. withstand the tensile forces exerted by the muscles involved in strenuous actions of the body wall, spine, hip and knee.

The planes of the articular surfaces of the sacroiliac joints face inwards, downwards and forwards. The sacrum is in effect suspended from the hip bones by the ligaments of the sacroiliac joints. These ligaments are accordingly very strong, particularly the interosseous sacroiliac ligament, which is probably the strongest ligament in the body. The sacrotuberous and sacrospinous ligaments are also strong and restrain the lower part of the sacrum from rotating upwards and backwards, about the pivot of the interosseous ligament.

When standing, weight is transmitted from the sacrum to the femora through each arm of a **femorosacral arch**, which comprises the sacrum, the sacroiliac joints, the arcuate lines of the pelvic brim and the lunate surfaces of the acetabula. Strong trajectories of trabeculae are developed along these pathways, and can be followed across the hip joints into the principal compressive groups of trabeculae in the upper ends of the femora. **When sitting**, weight is transmitted from the sacrum to the ischial tuberosities through each arm of an **ischiosacral arch**, which comprises the sacrum, the sacroiliac joints, the posterior parts of the ilia, the ischial bodies and tuberosities. Again, strong trajectories of trabeculae are developed to meet these loads. The extremities of the femorosacral and ischiosacral arches are coupled by subsidiary arches. The upper subsidiary arch, between the arms of the femorosacral arch, is formed on each side by the superior pubic ramus and the body of the pubis, joined at the symphysis. The component of this arch on each side is referred to as the **anterior column** of the acetabulum. The lower subsidiary arch, between the arms of the ischiosacral arch, is formed on each side by the ischiopubic ramus and body of the pubis, joined at the symphysis. The body of the ischium and the ischiopubic ramus is referred to as the **posterior column** of the acetabulum. The subsidiary arches resist both compression and distraction of the main arches. All four arches combine to form a solid bony ring, weakened only at the sacroiliac joints and pubic symphysis. Disruption of this ring by fracture or dislocation requires considerable force, and can be a serious, and sometimes fatal injury, due to the vascular and other soft tissue complications that can ensue.

The pull of most of the muscles attached to the pelvis is from bony prominences or surfaces with thick margins, within which stout trabecular systems are developed. A particularly strong group of trabeculae run from the tubercle of the crest of the ilium to the acetabulum, and withstands the abduction pull of the glutei. Nevertheless, even mature bone may not be strong enough to withstand the full force of some muscle contractions against resistance, and avulsion fractures at the sites of tendinous attachments occur. More commonly these injuries occur through the growth plates of secondary centres of ossification e.g. of the anterior iliac spines and the ischial tuberosity.

The following questions on the pelvis are concerned with the female, in relation to the function of childbearing. They refer to what might be called the obstetric pelvis. Average measurements only are given. There is obviously, however, some variation.

249. What is the pelvic inlet ?

The pelvic inlet is the superior aperture of the lesser pelvis bounded by the pelvic brim. It is usually slightly transversely oval in outline, with a small indentation posteriorly caused by the sacral promontory. The plane of the inlet makes an angle of forward inclination with the horizontal, when standing, of 50–60°. The sacral promontory in this position is 9–10 cm above the upper border of the symphysis pubis. A transverse plane at the level of the upper border of the symphysis in this position passes through the ischial spines, the tip of the coccyx, the centres of the heads of the femora and the tips of the greater trochanters. A line directed perpendicularly to the plane of the inlet, from its centre, called the axis of the inlet, projects downwards through the middle of the coccyx and upwards through the umbilicus. The anteroposterior diameter of the inlet, measured between the sacral promontory and the upper border of the pubic symphysis, is called the **true conjugate** and measures 11–11.5 cm. The oblique diameter, between a sacroiliac joint and the opposite iliopubic eminence, is about 12.5 cm. An oblique diameter is designated right or left according to the sacroiliac joint through which it passes. The maximum transverse measurement is about 13.5 cm. It is not strictly a diameter since it passes a little behind the centre of the inlet.

250. What is the pelvic outlet ?

The pelvic outlet is the inferior aperture of the lesser pelvis bounded by the coccyx, sacrotuberous ligaments, ischial tuberosities, ischiopubic rami and lower borders of the pubic bodies and symphysis pubis. It is rhomboidal in outline, with the long diameter directed sagittally. The plane of the outlet, between the coccyx and the lower border of the pubic symphysis, makes an angle of forward inclination with the horizontal, when standing, of about 15°. The ischial tuberosities descend below this plane, so that two triangular planes, anterior and posterior, can alternatively be described. A perpendicular line from the centre of the plane of the outlet, called the axis of the outlet, projects to the sacral promontory. The anteroposterior diameter of the outlet in obstetrics is measured between the inferior border of the symphysis and the apex of the sacrum, at the sacrococcygeal joint, and is 11.5–13.5 cm. During delivery the coccyx is displaced backwards, increasing tne theoretical anterior diameter of the outlet, measured from the tip of the coccyx, by about 2 cm. The transverse diameter, between the inner borders of the ischial tuberosities, is 11–11.5 cm.

251. What is the pelvic canal ?

The pelvic canal is the cavity of the lesser pelvis, between the superior and inferior apertures. The anterior wall is flat, about 4 cm long, and comprises only the width of the symphysis and the bodies of the pubic parts of the hip bones. The posterior wall is about 11 cm long and comprises the anterior surfaces of the sacrum and coccyx. It is usually flat above, but may have a more acute curve forwards at the level of the fourth segment. Each side wall is about 7.5 cm deep and is made up of the quadrangular aspects of the combined ischial and lower iliac parts of the hip bone, with the ischial spine curving towards the margins of the sacrum and coccyx. A curved line joining the centres of the planes of the inlet and outlet, parallel with the posterior wall of the cavity, is the axis of the cavity. It is a guide to the direction of pull required when obstetric forceps are applied.

Two measurements of the pelvic canal used in obstetrics are the diagonal and obstetric conjugates. The **diagonal conjugate** is the measurement between the lower border of the symphysis and the sacral promontory and is 1.5–2 cm greater than the true conjugate, which is 12.5–13.5 cm. The **obstetric conjugate** is the measurement from a point 1 cm down the posterior surface of the symphysis from its upper border, to the sacral promontory, and is approximately 0.5 cm less than the true conjugate, because of the obliquity of the symphysis. In obstetric terms, it is the measurement of the effective anteroposterior diameter of the pelvic brim.

The roomiest part of the canal is about halfway down, in a plane passing through the middle of the back of the symphysis and the lower border of the second sacral vertebra. The narrowest part of the canal is lower down, in a plane passing through the lower border of the symphysis, the ischial spines and the lower border of the fourth sacral vertebra. It is level with the uterovaginal angle. The bony diameters of the pelvic cavity are, of course, reduced by the soft tissues lining the walls, and by the contained viscera.

The ligaments of the sacroiliac joints, and of the symphysis soften a little during pregnancy, from the effect of relaxin. The resulting laxity of the joints can be demonstrated radiologically from the second month of pregnancy. Laxity of the symphysis, in particular, can be disconcerting.

252. What are the anatomical features of an ideal pelvis, from an obstetric point of view?

Ideally, from an obstetric point of view, the female pelvis should show many of the following characteristics. The **inlet** should be rounded or somewhat oval in the anteroposterior diameter, with no undue projection of the sacral promontory. In the erect position the plane of the brim should be at an angle of 50–60° to the horizontal. The **canal** should be shallow, with straight side walls from which the ischial spines do not project inwards unduly. The sacrum should show a smooth concave curve both longitudinally and transversely. The pubic arch of the **outlet** should be well curved, the inferior pubic rami forming an angle of 85–90°, to allow the fetal head to make full use of the available space as it extends under the arch during delivery. The transverse diameter between the ischial tuberosities should be wider than the 9.5 cm biparietal diameter of the average fetal head.

253. What different shapes of pelvic inlet are recognised?

Four different shapes of inlet are described, which modify the average measurements in individual subjects. The **gynaecoid** pelvis has a rounded rim with only shallow posterior indentation by the sacral promontory. It is present in about 40% of women. The **anthropoid** pelvis has an oval brim, with the long diameter directed antero-posteriorly. It is present in about 25% of women. The **android** pelvis has a wedge-shaped brim with the apex anteriorly, and is present in about 32% of women. The **platypelloid** or flat brim is flattened anteroposteriorly, and is present in about 3% of women. Obstetrically the gynaecoid and the anthropoid pelves are favourable to safe delivery, but platypelloid and android pelves can cause difficulty.

THE HIP BONE, PELVIC BONE OR OS INNOMINATUM

254. What is the anatomical position of the hip bone?

The articular surface of the pubis is in the sagittal plane. The anterior superior iliac spine is in the same coronal plane as the pubic tubercle. The spine of the ischium is in the same transverse plane as the pubic tubercle.

255. What are the three main parts of the hip bone?

The **ilium** comprises the upper, weight bearing, 35% of the acetabulum and the bone above it. The **pubic part** comprises the anterior 20% of the acetabulum, the superior pubic ramus anterior to it, the body of the pubis, and the inferior pubic ramus extending posteriorly to meet the ramus of the ischium below the middle of the obturator foramen. The **ischium** comprises the posterior 45% of the acetabulum, the body and the tuberosity below it, and the ramus projecting anteriorly to meet the ramus of the pubis.

256. Where is the fascia lata attached to the hip bone ?

Beginning at the anterior superior iliac spine it is attached along the outer lip of the iliac crest, enclosing the origin of tensor fasciae latae, to the junction with the crest of the posterior gluteal line. Here the fascia encloses the origin of gluteus maximus. The fascia rejoins the bone along the lower part of the medial margin of the ischial tuberosity, lateral to and below the insertion of the sacrotuberous ligament. The attachment then follows the lower border of the ischiopubic ramus, medial to the attachments of adductor magnus, gracilis and adductor longus, to below the pubic tubercle, where it fades laterally into the anterior epimysium of pectineus along the pecten pubis of the superior pubic ramus. Along the ischiopubic ramus it is lateral to the attachment of the membranous layer of the superficial fascia of the perineum, called Colles' fascia.

257. Identify the anterior superior iliac spine. What is attached to it ?

It is the prominence of the upper part of the anterior border of the ilium, where it joins the iliac crest. Anteriorly are attached the inguinal ligament, fascia lata and sartorius. Superiorly are attached fibres of external oblique, internal oblique and transversus abdominal muscles. Laterally are attached fibres of tensor fasciae latae and gluteus minimus, and medially fibres of iliacus.

258. Identify the anterior inferior iliac spine. What is attached to it ?

It is the prominence of the lower part of the anterior border of the ilium. To the upper part is attached the straight head of rectus femoris. The iliofemoral ligament is attached to the lower part.

259. Identify the tubercle of the crest of the ilium. What is attached to it ?

It is the prominence of the outer lip of the iliac crest at the junction of the anterior and middle thirds of the ventral segment, about 5 cm behind the anterior superior spine. To it is attached the part of the fascia lata forming the upper part of the iliotibial tract behind the origin of tensor fasciae latae over gluteus medius.

260. How can prominences of the hip bone be used to determine levels of the lower spine ?

The **supracristal plane** is at the level of the upper limits of the iliac crests and passes through the spine of the fourth lumbar vertebra. The **intertubercular plane** is at the level of the tubercles of the iliac crests which can be felt by running the fingers along the crests backwards from the anterior superior iliac spines. It passes through the spine of the fifth lumbar vertebra.

These planes are useful to ascertain the correct level for spinal tap or injection. The **interspinous plane**, at the level of the anterior superior iliac spines, and of the 'dimples of Venus' over the posterior superior iliac spines, passes through the promontory of the sacrum and the spine of the second sacral vertebra. The dural sac ends at this plane.

261. What structures are attached to the iliac crest ?

The iliac crest consists of a ventral segment made up of the anterior two thirds and a dorsal segment made up of the posterior third. The **ventral segment** has an outer lip and an inner lip, enclosing an intermediate area. To the outer lip of the ventral segment is attached the fascia lata to its whole extent, tensor fasciae latae from the anterior superior spine to just in front of the tubercle of the crest, external oblique to its anterior two thirds, and latissimus dorsi to its posterior part. There is a variable interval between the posterior border of the external oblique and the lateral border of latissimus dorsi, which forms the inferior border of the **lumbar triangle of Petit**. To the anterior three quarters of the inner lip is attached transversus abdominis and to the posterior quarter quadratus lumborum. To the whole length of the intermediate area is attached internal oblique. The **dorsal segment** has sloping outer and inner surfaces. To the outer surface is attached gluteus maximus enclosed in fascia lata. To the inner surface is attached the iliolumbar ligament anteromedially, erector spinae posterolaterally, and the posterior sacroiliac ligament medially.

262. What nerves would be cut in an incision along the iliac crest ?

An incision along the iliac crest, for example to obtain a bone graft, and depending on its extent, would cut from behind forwards, the lateral branches of the upper three lumbar dorsal rami, the lateral cutaneous branch of the iliohypogastric nerve, the lateral cutaneous branch of the subcostal nerve, and the lateral cutaneous nerve of the thigh, just in front of the anterior superior iliac spine.

263. From where on the hip bone do muscles arise that flex the hip joint ?

Iliacus arises from about the upper two thirds of the iliac fossa. Inferiorly, towards the groove between the anterior iliac spine and the iliopubic eminence, for iliopsoas, there is a bursa separating this muscle from the bone. It communicates with the hip joint.

    Rectus femoris arises by two tendinous heads. The straight head is attached to the upper part of the anterior inferior iliac spine. The reflected head is attached to a groove above the acetabulum.

    Sartorius arises by tendinous fibres from the anterior superior iliac spine and the upper half of the notch below it.

Pectineus arises from the pecten pubis and the bone anterior to it, between the pubic tubercle and the iliopubic eminence.

264. From where on the hip bone do muscles arise that adduct the hip joint ?

Adductor longus arises from the femoral surface of the body of the pubis in the angle between the crest and the symphysis.

Gracilis arises from the lower medial part of the femoral surface of the body of the pubis and along the lower margin of the inferior pubic ramus and part of the ramus of the ischium.

Adductor brevis arises from a narrow attachment to the femoral surfaces of the body of the pubis below adductor longus and the inferior pubic ramus between obturator externus and gracilis. Further laterally it arises for a short distance between adductor magnus and gracilis.

The adducting fibres of **adductor magnus** arise from the femoral surface of the posterior part of the inferior ramus of the pubis, between obturator externus and adductor brevis, and more posteriorly from the ramus of the ischium between obturator externus and gracilis. The most posterior fibres from the lateral part of the lower half of the ischial tuberosity extend rather than adduct the hip joint.

265. From where on the hip bone do muscles arise that extend the hip joint ?

Part of **gluteus maximus** arises from the gluteal surface of the ilium behind the posterior gluteal line. The hamstring tendons arise from the tuberosity of the ischium.

Semimembranosus arises from the upper outer portion of the upper part of the tuberosity.

Semitendinosus and the long head of **biceps** arise together from the lower inner portion of the upper part of the tuberosity.

The hamstring part of **adductor magnus** arises by tendinous fibres from the outer portion of the lower part of the tuberosity.

266. From where on the hip bone do muscles arise that abduct the hip joint ?

Gluteus medius arises from the gluteal surface of the ilium bounded by the posterior and anterior gluteal lines, the outer lip of the iliac crest and the greater sciatic notch.

Gluteus minimus arises from the same surface between the anterior and inferior gluteal lines, the anterior border of the ilium and the greater sciatic notch.

Tensor fasciae latae arises from the outer lip of the iliac crest from the anterior superior iliac spine to the tubercle of the iliac crest.

Sartorius assists in abduction. It is attached to the anterior superior iliac spine and the upper half of the notch below it.

267. From where on the hip bone do muscles arise that laterally rotate the hip joint ?

Obturator externus arises from the femoral surface of the obturator membrane and the adjacent outer surfaces of the pubic and ischial rami.

Obturator internus arises from the pelvic surface of the obturator membrane, the adjacent pelvic surfaces of the inferior pubic ramus and ramus of the ischium, and the pelvic surface of the bone from the obturator foramen anteroinferiorly to the greater sciatic notch posterosuperiorly.

Gemellus superior arises from the gluteal surface of the ischial spine.

Gemellus inferior arises from the tuberosity of the ischium at the lower end of the lesser sciatic notch, below the groove for the tendon of obturator internus.

Quadratus femoris arises from the upper lateral aspect of the tuberosity of the ischium.

Sartorius and gluteus maximus assist in lateral rotation of the hip joint (see questions 263 and 265).

268. From where on the hip bone do muscles arise that medially rotate the hip joint ?

Medial rotation is produced by the anterior fibres of gluteus medius and minimus, and sartorius, usually assisted by the adductors (see questions 263, 264 and 266).

269. What channels lead out from the side-wall of the pelvis ?

Three channels lead out from the side-wall of the pelvis — the greater sciatic foramen, the lesser sciatic foramen and the obturator canal. The greater sciatic foramen and the obturator canal are possible sites of herniation of small bowel.

270. What are the boundaries of the greater sciatic foramen ?

The greater sciatic foramen is bounded by —

    i. the **greater sciatic notch** of the hip bone.

    ii. the **sacrotuberous ligament**, arising from the posterior superior iliac spine, the posterior inferior iliac spine, and the smooth area of the gluteal surface of the ilium in front of the lower spine, towards the posterior gluteal line. It also arises from the third and fourth transverse tubercles of the sacrum, the lateral margin of the sacrum below tne sacroiliac joint, and the lateral margin of the coccyx. Upper fibres merge with the dorsal sacroiliac ligament. It is inserted into the medial margin of the ischial tuberosity and forms a falciform projection of fibres along the lower border of the ramus of the ischium.

iii. the **sacrospinous ligament**, attached to the spine of the ischium, with the coccygeus muscle lying in front of it. It arises from the lateral margins of the lower part of the sacrum, and the coccyx. It is anterior to the sacrotuberous ligament.

271. What passes through the greater sciatic foramen ?

The greater sciatic foramen is mostly occupied by the emerging piriformis. Above the muscle pass the superior gluteal vessels and nerves. Below the muscle pass the inferior gluteal vessels and nerve; the sciatic nerve with the nerve to quadratus femoris deep to it and the posterior cutaneous nerve of the thigh superficial to it; the nerve to obturator internus over the base of the spine of the ischium; the internal pudendal vessels over the apex of the spine of the ischium; and the pudendal nerve over the sacrospinous ligament. Clinically two apertures of the foramen are described, the suprapiriformic and the infrapiriformic, through which small bowel can occasionally herniate.

272. What are the surface markings of the greater sciatic foramen ?

The important part of the greater sciatic foramen is where the sciatic nerve lies against its inferior margin. This point lies roughly at the level of intersection of a line drawn horizontally between the tips of the greater trochanters and a line drawn vertically through the medial side of the ischial tuberosity. The upper border of the piriformis, is along a line drawn between the tip of the greater trochanter and the dimple caused by the posterior superior iliac spine. The emerging sciatic nerve lies about 3 cm below and medial to the midpoint of this line.

273. What are the boundaries of the lesser sciatic foramen ?

They are the lesser sciatic notch and the sacrotuberous and sacrospinous ligaments.

274. What passes through the lesser sciatic foramen ?

The surface of the bone is covered with a layer of fibrocartilage, slightly grooved to correspond to the fasciculi of the tendon of obturator internus emerging through the foramen and running over it. The tendon is joined by the two gemelli muscles arising from the margins of the foramen above and below the tendon. Above the tendon pass the nerve to obturator internus on the base of the ischial spine, the internal pudendal vessels on the apex of the spine at the point of attachment of the sacrospinous ligament, and the pudendal nerve in contact with the sacrospinous ligament. These three structures pass out of the greater sciatic foramen and then through the lesser sciatic foramen to reach the lateral wall of the ischiorectal fossa.

275. What is the obturator canal ?

The obturator canal is the passage, about 2–3 cm long, leading from the pelvic cavity to the adductor region of the thigh. It is bounded by the obturator groove of the superior pubic ramus, the upper border of the obturator membrane, and the upper borders of obturator internus and externus. It contains the obturator nerve and vessels. The nerve is usually placed above the vessels. A hernial sac can follow either division of the nerve. The sac can compress the nerve to produce hip and knee pain in about half the patients who present with this rare condition.

276. What part of the acetabulum is covered by articular cartilage ?

It is the lunate surface of the acetabulum that is covered with articular cartilage. This is the obvious smooth horse-shoe shaped surface within the cup, widest superiorly, where the cartilage is thickest and where most weight bearing stress is borne in the erect position. It surrounds the acetabular fossa in front, above and behind.

277. What structures could be damaged by orthopaedic instruments or fixation devices inadvertently pushed through the floor of the acetabulum ?

The structures damaged in such an event will depend on the part of the acetabulum pierced and the direction of penetration. If penetration is directed **posteriorly** the sciatic nerve can be damaged, as can also the other nerves and vessels emerging from the greater sciatic foramen below piriformis, namely the posterior cutaneous nerve of the thigh, the nerve to quadratus femoris, the inferior gluteal vessels and nerve, the nerve to obturator internus, internal pudendal vessels and the pudendal nerve. If penetration is directed **superiorly** the iliopsoas will be entered and the external iliac vessels can be damaged. The common iliac vein particularly lies directly on the bone of the brim for a short distance. If penetration is directed **medially** a variety of structures can be involved. Obturator internus is only about 3–4 mm thick. Medial to the muscle, lying against the obturator fascia is the profuse paravesical plexus of veins, together with the obturator nerve, and the obturator vessels, the superior and inferior vesical vessels, the middle rectal vessels, and the uterine vessels, all arising from the internal iliac vessels. Also close to are the ureter, the ovary in the ovarian fossa, the bladder, uterus and rectum.

278. Where are the ligaments of the hip joint attached to the hip bone ?

The **iliofemoral ligament** is attached to the lower part of the anterior inferior iliac spine. The **pubofemoral ligament** is attached to the lateral end of the obturator crest and the adjoining rim of the acetabulum. The **ischiofemoral ligament** is attached to the body of the ischium below and behind the rim of the acetabulum. The **ligamentum teres** is attached to the anterior and posterior obturator tubercles and the transverse acetabular ligament between them, across the acetabular notch.

279. How does the acetabulum develop around the head of the femur ?

At birth the acetabulum consists of cartilage. During infancy the **primary centres** of ossification of the ilium, ischium and pubis converge to mould the cartilaginous acetabulum into a cup directed laterally and a trifid stem directed medially. At about 8 years of age **secondary centres** of ossification appear in the cup. The largest centre is for the pubic portion of the cup, forming the anterior wall of the acetabulum. It is called the os acetabuli. The centre for the iliac portion of the cup forms a large part of the superior wall of the acetabulum. The centre for the ischial portion of the cup is small, and is rarely seen on X-ray examination. The centres of ossification all fuse by the age of about 18 to 19 years of age. They can be confused with avulsion fractures or loose bodies within the hip joint. The proper development of the acetabulum requires the presence of the head of the femur in the acetabulum. In untreated congenital dislocation of the hip the acetabulum remains hypoplastic.

280. How are the pubic bones joined together ?

They are joined by the pubic symphysis, a fibrocartilaginous joint. The uneven articular surfaces, each covered by a layer of hyaline cartilage, are joined to each other by a lamina of fibrocartilage. After the age of about 10 years the fibrocartilage cavitates to a varying degree downwards and for-wards from the upper posterior part of the joint. The joint is reinforced by a superior ligament reaching between the pubic tubercles, and by an inferior arcuate ligament, at the apex of the pubic arch. Crossed fibres of the medial heads of origin of rectus abdominis reinforce the joint anteriorly, making this the strongest support.

281. What is attached to the pubic crest ?

The lateral origin of rectus abdominis arises from the pubic crest. Transversalis fascia covers its posterior part. Pyramidalis arises from the body of the pubis in front of it and behind the anterior sheath of the rectus.

282. How is the inguinal ligament attached to the hip bone ?

The inguinal ligament (Poupart's ligament) results from a backward folding of the lower border of the aponeurosis of external oblique. It is the thickest and strongest part of the aponeurosis. Narrow laterally, it is attached to the anterior superior iliac spine. As the fibres of the aponeurosis descend medially to the ligament they enter it at an angle of about 10–20°, so that fibres from a wide area can be closely packed together. Medially the ligament widens to a band between 1.5 and 2 cm wide. The anterior, superficial fibres attach to the pubic tubercle forming the lateral crus of the superficial inguinal ring, curved above to form a groove for the spermatic cord. The posterior, deeper fibres spread out posteriorly and posterolaterally to form a triangular horizontal membrane attached to the medial end of the pecten pubis. This part is called the lacunar ligament (Gimbernat's ligament). Its concave lateral border forms the medial rim of the femoral ring. It extends laterally as a thin but strong fibrous band attached to the lateral part of the pecten pubis as far as the iliopubic eminence. It fuses with the upper attachment of the anterior fascia of pectineus to the pecten pubis. This extension of the lacunar ligament is called the pectineal ligament, or ligament of Astley Cooper. By passing down in front of the pectineal ligament the inguinal ligament allows space for the spermatic cord and large vessels of the lower limb to leave the abdomen.

283. What is attached to the pecten pubis ?

The pecten pubis is the sharp edge of the linea terminalis, which extends between the pubic tubercle and the iliopubic eminence. Attached to it are the anterior fascia of pectineus, the fibrous extension of the lacunar ligament known as the pectineal ligament or the ligament of Astley Cooper, the lateral expansion of the rectus abdominis sheath known as the ligament of Henle medially, and the conjoint tendon laterally. Transversalis fascia reaches it posteriorly.

284. What structures are in contact with the pelvic surface of the body of the pubis ?

From the lower aspects of the posterior surfaces of the bodies of the pubes, on either side of the median plane, arise stout bands of fibromuscular tissue called the **puboprostatic ligaments** in the male, and the **pubovesical ligaments** in the female. Between them is a midline hiatus through which passes the deep vein of the penis and other small veins. Lateral to the origin of each puboprostatic or pubovesical ligament arise the anterior fibres of the **levator ani**. Above these ligaments the prostate and base of bladder, surrounded by a venous plexus, are separated from the pubic bones by the retropubic space (the **cave of Retzius**). This space contains fat covered above by peritoneum, raised in the midline by the urachus, to form the median umbilical ligament, at the anterior limit of the paravesical fossae of the pelvic cavity.

285. Which part of the bone takes the weight of the trunk in the sitting position ?

The medial surface of the ischial tuberosity below the origin of the hamstring muscles is the bone which takes the weight when sitting. The surface of the bone is covered by a bursa and fibro-fatty tissue. The gluteus maximus overlies it on standing.

286. Where is the pelvic diaphragm attached to the hip bone ?

The pelvic diaphragm consists of the levator ani and coccygeus muscles with their covering superior and inferior fasciae. The levator ani is attached to the hip bone in a line sloping slightly upwards across the lower lateral part of the posterior aspect of the body of the pubis, and the fascia covering the inner surface of obturator internus, to the spine of the ischium. Coccygeus arises from the tip of the spine of the ischium, anterior to and coextensive with the sacrospinous ligament, and overlapped anteriorly a little on its deep surface by levator ani.

287. Where is the urogenital diaphragm attached to the hip bone ?

The urogenital diaphragm consists of the sphincter urethrae and deep transverse perinei muscles, enclosed in fascia. The muscles occupy the deep perineal pouch. The inferior layer of fascia is strong and comprises the **perineal membrane**. The anterior limit of the membrane forms the transverse perineal ligament separated from the symphysis and inferior pubic ligament by a narrow interval containing the deep dorsal vein of the penis. The perineal membrane is attached to the summit of the convex inner surface of the adjoining pubic and ischial rami separating the perineal from the pelvic surfaces of this part of the bone. The muscles and the superior layer of fascia are attached above the perineal membrane.

288. Where is the surface of the bone which forms the lateral limit of the superficial perineal pouch and what is in contact with it ?

The superficial perineal pouch is inferior to the perineal membrane, between it and the membranous layer of the superficial fascia (the fascia of Colles). The two layers meet posteriorly, enclosing the superficial transverse perineal muscles. The membranous layer of the superficial fascia is attached to the inferior border of the pubic ramus and the adjoining part of the ramus of the ischium. The perineal membrane is attached above this line to the convexity of the inner surfaces of the adjoining pubic and ischial rami. Between these two layers, against the bone, is attached a crus of the penis or clitoris, covered by the ischiocavernosus muscle.

289. Where is the ischiorectal fossa in relation to the bone ?

It is medial to the body and the posterior part of the ramus of the ischium, below the attachment of the pelvic diaphragm, and above the attachment of the sacrotuberous ligament. It is posterior to the posterior attachment of the perineal membrane. It should be called the ischio-anal fossa since it is the anal canal, and not the rectum that forms its medial boundary.

290. What nerves are in close contact with the hip bone ?

The sciatic nerve lies against the posterior aspect of the acetabulum. The obturator nerve passes through the obturator canal. The femoral nerve is separated from the ilium by the iliopsoas. The posterior primary rami of the upper three lumbar spinal nerves, the lateral cutaneous branch of the ilio-hypogastric nerve, the lateral cutaneous branch of the subcostal nerve, and the lateral cutaneous nerve of the thigh are in relation to the crest of the ilium. The superior gluteal nerve is in contact with the superior border of the greater sciatic notch. The inferior gluteal nerve, the nerve to quadratus femoris, the sciatic nerve, the posterior cutaneous nerve of the thigh, and the nerve to obturator internus are in relation to the inferior margin of the greater sciatic notch. The nerve to obturator internus curves over the posterior surface of the spine of the ischium to pass forwards below its inferior border. The pudendal nerve takes a similar course to the nerve to obturator internus, but over the sacrospinous ligament, just medial to the tip of the ischial spine. The perineal branch of the pudendal nerve and the dorsal nerve of the penis are in relation to the inferior pubic ramus, in the deep perineal pouch. The accessory obturator nerve passes in front of the superior pubic ramus to deep to the pectineus muscle.

291. What blood vessels could be torn if the hip bone is fractured ?

Most bleeding, which can be massive, from fractures of the pelvis is thought to be venous in origin. The vessels involved may be marrow vessels, adjacent parietal vessels and in severe cases, more remote visceral vessels. **Marrow vessels** include branches of iliolumbar, superior and inferior gluteal and deep circumflex iliac vessels which supply the ilium; and of obturator, and medial and lateral circumflex femoral vessels which supply the ischium, acetabulum and pubis. Adjacent **parietal vessels** include common, external and internal iliac vessels and branches of internal iliac vessels. Of these last, the superior gluteal vessels are especially vulnerable on account of their kinked course around the margin of the greater sciatic notch. Other parietal branches of the internal iliac vessels are the obturator, inferior gluteal, iliolumbar and pudendal. The more remote, **visceral vessels** include the paravesical plexus of veins, and superior vesical, inferior vesical, middle rectal, uterine and vaginal vessels.

292. What secondary centres of ossification are there in the hip bone ?

Secondary centres of ossification of the hip bone appear between puberty and the early teens. Usually two centres, which fuse quickly, appear for the iliac crest. The anterior superior iliac spine may develop from an ossification centre separate from that of the crest. It can be avulsed by a strong pull of sartorius against resistance. A centre for the anterior inferior iliac spine can similarly be avulsed, less commonly, by traction of the straight head of rectus femoris. A centre for the ischial tuberosity may be the last to unite. It grows forward from behind, fusing with the ischiopubic ramus as it does so. Part of it may be avulsed by action of the hamstrings. Pubic centres sometimes appear in the pubic tubercle, pubic crest, and the articular surface of the symphysis. Acetabular centres appear in the pubic, iliac, and sometimes the ischial parts of the cartilaginous cup.

THE FEMUR

293. How are the neck and head of the femur set on the shaft ?

The neck and head of the femur curve upwards, medially and a little anteriorly from the shaft; hence the description of two angles —

i. the **angle of inclination**, between the longitudinal axis of the neck and head of the femur and the axis of the shaft. It is normally about 125°, with a range between 110° and 140°. It may be more, called **coxa valga**, if the hip develops in the absence of the normal pull of the gluteal muscles (e.g. poliomyelitis or spastic paralysis), or it may be less, called **coxa vara**, as in rickets, when the softer bone is deformed by the weight of the body.

ii. the **angle of version** or **torsion**, between the longitudinal axis of the neck and head of the femur and the transverse line between the condyles, when viewed along the shaft from below. It is about 30° at birth, 25° at 5 years, 20° at 10 years, and 15° at 15 years. In the adult it is about 12°. Being anteriorly directed, the angle is described as one of anteversion, in contrast to the situation in the humerus, where the corresponding angle is one of retroversion. The angle may be more than normal at birth, up to 60°, as occurs in association with congenital dislocation of the hip. Failure of a wide angle of version to revert to normal with growth in childhood results in a persistent intoeing gait. There is an increased range of medial rotation and a corresponding decreased range of lateral rotation of the hips.

294. Where is the capsule of the hip joint attached to the femur ?

The capsule of the hip joint is attached anteriorly to the intertrochanteric line. From the upper tubercle of the intertrochanteric line the capsular attachment passes backwards over the lateral part or base of the neck, medial to the greater trochanter and the insertions of obturator internus and the gemelli, obturator externus, and piriformis, and just lateral to, and so enclosing, a short portion of the upper epiphyseal line of the greater trochanter. Posteriorly the attachment passes medially and downwards about 1 cm medial to the intertrochanteric crest, and then curves round the under surface of the neck above and in front of the lesser trochanter, to reach the inferior tubercle of the intertrochanteric line.

295. Where are the ligaments of the hip joint attached ?

The ligaments of the hip joint are thickenings of the capsule. They spiral distally from their proximal attachments such that extension of the femur, with abduction and medial rotation, tightens the joint into its close-packed position.

The **iliofemoral ligament**, or Y-shaped ligament of Bigelow, passes from the lower part of the anterior inferior iliac spine, downwards and laterally, to spiral into the intertrochanteric line. It has thickened upper and lower bands, and a thinner middle part. It covers the front of the head of the femur except for its lower medial part. It vies with the interosseous sacroiliac ligament for the status of the strongest ligament of the body, and is not often torn when the hip is dislocated. Its remarkable strength is attributed to its role in resisting the force of body weight which, in the erect posture, passes posterior to a line joining the heads of the femora, and acts to extend the relaxed joint.

The **pubofemoral ligament** passes from the iliopectineal eminence and the lateral end of the obturator crest, downwards and backwards over the lower medial aspect of the head of the femur, to the lower aspect of the base of the neck.

The capsule between the lower margin of the iliofemoral ligament and the upper part of the pubofemoral ligament is thin and may communicate (in 15% of subjects) with the psoas bursa.

The **ischiofemoral ligament** passes from the posteroinferior aspect of the acetabulum, near its margin, to the posterosuperior part of the base of the neck. It is the thinnest ligament. The upper fibres are horizontal, the lower fibres nearly vertical.

The capsule is thin between the ischiofemoral ligament and the more lateral zona orbicularis.

The **zona orbicularis** consists of a group of deep fibres in the capsule, seen only convincingly posteriorly, which form part of a collar constricting the lateral part of the capsule. It does not attach to bone.

The **ligamentum teres** or round ligament of the head of the femur (though it is flat rather than round) arises from the transverse acetabular ligament and its bony attachments, continuous with the labrum. It is reinforced by other fibres from the margins of the acetabular fossa. Ensheathed by synovium, it is inserted into the fovea capitis. It is probably not stressed by any physiological movement. It carries a variable blood supply to the head of the femur, from the obturator artery and/or medial circumflex branch of the profunda artery.

296. How does the hip joint contrast with the shoulder ?

i. The **head** of the femur constitutes more than a hemisphere and is deeply set in the acetabulum, so that the stability of the hip joint is contributed to considerably by bony contours. The head of the humerus constitutes less than a hemisphere and is set in a shallow glenoid fossa a third of its size, so that little stability is contributed to the shoulder joint by bony contours.

ii. The free edge of the **labrum** of the acetabulum has a smaller circumference than its attached base, so that it helps to retain the head in position. Dislocation of the head tears the labrum. The glenoid labrum contributes little to the stability of the shoulder joint.

iii. The **neck** of the femur is longer and narrower than the head, allowing more movement than would otherwise be possible. The hip joint is second only to the shoulder in its range of mobility. The anatomical neck of the humerus is barely discernible.

iv. The **capsule** of the hip joint is reinforced by strong ligaments, but not by tendon. The capsular ligaments of the shoulder joint are relatively weak and the capsule is lax. The lax capsule is, however, considerably strengthened by fusion with the tendons of the rotator cuff of muscles.

v. The **muscles** immediately surrounding the hip joint insert some distance from the head of the femur, and so can exert relatively more leverage on the femur than those of the rotator cuff on the humerus.

vi. The **ligamentum teres** carries some blood supply to the head of the femur. There is no such ligament in the shoulder joint.

297. What is the nerve supply of the hip joint ?

The hip joint receives branches from —

i. the **obturator nerve**, from the main trunk in the pelvis or obturator canal, or from either its anterior or posterior branch.

ii. the **accessory obturator nerve**, small and present in only about one in ten subjects. It crosses the superior pubic ramus to the deep surface of pectineus, from where the articular branch arises.

The obturator and accessory obturator nerves supply the area of the pubofemoral ligament.

iii. the **femoral nerve**, from a muscular branch to the proximal posterior surface of rectus femoris. It supplies the region of the iliofemoral ligament, the pubofemoral ligament and the posterosuperior capsule.

iv. the **nerve to quadratus femoris**, which arises as the nerve passes through the greater sciatic foramen on the ischium deep to the sciatic nerve, obturator internus and gemelli, and quadratus femoris. It supplies the posterior part of the capsule.

v. the **superior gluteal nerve**. Its branches run between gluteus medius and minimus, and supply the lateral and superior aspect of the capsule.

Some relief of pain from osteoarthritis of the joint has been found to follow section of the obturator nerve and the nerve to quadratus femoris.

298. How does the upper end of the femur develop ?

At birth the head, neck and greater trochanter are cartilaginous. The advancing front of ossification from the shaft is upwardly convex, extending from below the greater trochanter to above the lesser trochanter. The lesser trochanter is cartilaginous, but is isolated from the head and neck by metaphyseal bone. A centre of ossification soon develops in the superolateral part of the head. It may appear as early as the third month. Most capital epiphyses (90%) are visible radiologically within $6\frac{1}{2}$ months (200 days) of birth. The form of the convex growth plate changes to that of an inverted cone, as it progresses proximally into the cartilaginous model. As seen on anteroposterior X-ray films, the medial limb of the inverted V of the growth plate is the shorter and advances into the neck at a greater rate of growth than the lateral limb which is the longer and lies opposite the greater trochanter and lateral aspect of the upper border of the neck. By the age of 4 the enlarging capital epiphysis, in the upper part of the head, lies above a horizontal growth plate, which includes the lower medial part of the head in the metaphysis. At this age also the centre for the greater trochanter appears. The capital epiphysis grows medially to form a gently hollowed out cup on the summit of the neck. The superolateral border of the neck remains cartilaginous until about 13 years of age when processes of ossification from the head and greater trochanter meet. Abnormal differential growth of the cervical and trochanteric aspects of the epiphyseal plate can result in varus or valgus deformity of the neck. The centre of ossification of the lesser trochanter appears at about the age of 13 years. The epiphyses are fused by 18. The neck of the femur is prone to slip on the shallow concavity of the adolescent capital epiphysis, so that the femur takes up a position of adduction and lateral rotation on the epiphysis. The lesser trochanteric epiphysis is prone to avulsion by the psoas, if sudden strong flexion of the hip is resisted.

299. What is the blood supply of the head of the femur in the adult ?

Blood reaches the head of the femur from three sources.

i. The principal supply is via the **retinacular vessels**. At the base of the neck an extracapsular anastomotic ring is formed by the ascending branches of the medial and lateral circumflex arteries, and by minor contributions from the superior and inferior gluteal arteries. The ascending branch of the medial circumflex artery is the largest component, the ascending branch of the lateral circumflex femoral artery being relatively small. The anastomotic ring may be incomplete. From the ring branches arise which pierce the lateral reaches of the capsule, to pass proximally under the synovium covering the neck, in fibrous prolongations of the capsule, called retinacula. Two principal groups of vessels are described, the superior and inferior, though some anterior and posterior vessels are present. They enter the bone near the articular cartilage, and curve into the head to form superoinferior arches between about 0.5 and 1 cm from the centre of the head and nearly 2 cm from the articular surface. From these arches smaller branches feed the periphery of the head, to the subchondral zone, especially of the upper quadrant.

ii. The **artery of the ligamentum teres** is a branch of the obturator, in the obturator canal, and or the medial circumflex branch of the profunda. It is variably developed. It may be absent. It may be a narrow vessel. It may be of significant size in a third of subjects, when after piercing the fovea capitis, it may anastomose with the branches of one or both of the retinacular groups of vessels, and probably then supplies a substantial portion of the head of the femur. Most commonly it supplies only a small area of the head adjacent to the fovea.

iii. Some minor **metaphyseal vessels** in the neck may reach the head. They arise from metaphyseal branches of the anastomotic ring, and from proximally directed branches of the nutrient artery or arteries in the medullary cavity.

In a displaced subcapital fracture of the neck of the femur the important retinacular vessels will be torn. They could also be occluded by a tense joint effusion. Only in a few subjects is the supply of blood through the ligamentum teres sufficient to maintain the vascularity of the whole head. Necrosis of the bone of the head is a common result of such fractures.

300. How does the blood supply of the head of the femur in a child differ from that in an adult ?

Up to the age of about 3 years the mainly cartilaginous head of the femur is fed predominantly by metaphyseal branches from the anastomotic ring around the base of the neck, formed between the ascending branches of the medial and lateral circumflex femoral arteries. At first these vessels pass directly through the metaphysis. As the epiphyseal plate matures, however, the metaphyseal supply is cut off and the capital epiphysis then depends on branches of the anastomotic ring which ascend the neck, bypassing the plate around its periphery. At first the ascending peripheral vessels branch from all round the anastomotic ring, which is supplied equally by lateral and medial circumflex arteries. By about the age of 5 years, however, the lateral circumflex contribution to the anastomotic ring is less prominent, and the ascending branches to the capital epiphysis become localised to upper and lower groups, each lying a little posteriorly, arising from a more developed ascending branch of the medial circumflex artery. Although present at birth the artery of the ligamentum teres quickly regresses and then does not make any contribution to the blood supply of the capital epiphysis until about the age of 8 years. The period between about the age of 5 and 10 years is that in which avascular necrosis of part or the whole of the capital epiphysis, Perthes' disease, can occur. It is probable that the period between the closure of the supply from the metaphyseal branches and the concentration of supply through groups of vessels from the ascending branch of the medial circumflex artery, and the reappearance of the artery of the ligament teres, is one of vulnerability. An episode of occlusion of the medial circumflex branch during this period, perhaps as it passes behind the iliopsoas, has been suggested as a possible cause of this disease.

301. What order of force is transmitted through the hip joint ?

The centre of gravity of the body, in the erect position, is considered to be just in front of the second segment of the sacrum. When standing on one leg its hip joint acts as the fulcrum of a Class 1 lever, one arm of which is acted on by the downthrust of the weight of the trunk and opposite leg, the other by the pull of the abductor muscles attached to the greater trochanter. In balance the moments about the fulcrum are equal. The distance between the centre of gravity of the body and the centre of the head of the femur is some two and a half times the distance between the centre of the head of the femur and the attachment of the abductor muscles to the greater trochanter. The head of the femur therefore transmits, in the stance phase of gait, not only the weight of the body less the supporting leg, but also a force two and a half times as great to counteract it, making a total downward thrust through the hip joint of about three times the weight of the body. When lifting, running or jumping, the force transmitted through the hip joint can be up to ten times the body weight. Lifting the straight leg from the supine position has been found to produce a force of one and a half times body weight through the hip joint. Such forces need to be taken into account in the design and fixation of replacement prostheses. They also explain the gait of a patient trying to ease the discomfort of a painful hip. By flexing the trunk laterally, over the painful hip on weight-bearing through it, the centre of gravity of the body is moved towards it, shortening the lever arm to the centre of the head of the femur.

302. What is the anatomical basis of the Trendelenberg test ?

When the normal subject stands on one leg only, the opposite side of the pelvis stays level or is raised by the action of the abductor muscles of the weight-bearing hip joint. A positive Trendelenberg sign is said to be present if, in these circumstances, the non weight-bearing side of the pelvis drops. This can occur if the abductor muscles of the weight-bearing hip are weak, as for instance after an attack of poliomyelitis, or if the weight-bearing hip joint is defective, as for instance in congenital dislocation of the hip, destruction of the joint by tuberculosis, old fracture of the neck of the femur, or after surgical excision of the joint (excision arthroplasty) for tuberculosis or osteoarthritis.

303. How are the trabeculae in the head and neck of the femur arranged ?

The complex three-dimensional structure formed by the trabeculae of the head and neck of the femur is revealed as a two-dimensional image on an X-ray film. The pattern of the trabeculae, seen as lines of various thickness, becomes evident after the infant begins to stand and walk. As in bone generally, the trabeculae follow the lines of compressive and tensile stresses.

The normal trabecular pattern, as seen on an anteroposterior X-ray film, is made up of five groups —

i. the **principal compressive group** of trabeculae radiate upwards in slightly curved lines from the medial cortex of the neck to the upper part of the head. They are thick and closely packed.

ii. the **secondary compressive group** arises from the medial cortex of the upper shaft and curves upwards and laterally towards the upper part of the neck and greater trochanter. These trabeculae are relatively thin and widely spaced.

iii. the **principal tensile group** extends from the lateral cortex just below the greater trochanter and curves upwards and inwards across the upper part of the neck to the lower part of the head. They are the thickest of the tensile trabeculae.

iv. the **secondary tensile group** extends from the lateral cortex below the principal tensile group, upwards and medially, to merge with the secondary compressive group.

v. the **trochanteric tensile group** extends from the lateral cortex just below the greater trochanter upwards to the upper border of the trochanter. They are slender and poorly defined.

The pattern made by these groups of trabeculae can be used to assess the degree of generalised osteoporosis. As the total bony mass lessens, the trochanteric and secondary compressive trabeculae are the first to disappear. The tensile trabeculae are the next to disappear, leaving only a much reduced number of principal compressive trabeculae. Prominent trabeculae can also be used to check the alignment of a fracture of the neck of the femur.

The principal compressive, the secondary compressive, and the principal tensile trabeculae outline an area in the neck of the femur containing thin trabeculae only, which can look empty as osteoporosis progresses. It is called **Ward's triangle**.

The **calcar femorale** is a vertical strut of dense cortical bone, visible only from within the marrow cavity, in the posteromedial part of the upper shaft, which extends from behind the lesser trochanter, upwards into the posterior part of the bone of the neck. It is thickest medially, thinning laterally. It can only be seen on a lateral X-ray film. It is often confused with, or mistakenly described as, the medial cortex of the neck of the femur.

304. What bursae are there near the hip joint ?

The more obvious bursae around the hip joint include —

    i. **anteriorly**, the bursa under the psoas tendon, which is said to communicate with the synovial cavity of the joint through the capsule in 15% of subjects.

    ii. **laterally**, three. There is one between the tendon of gluteus minimus and the anterior part of the lateral surface of the greater trochanter. There is one between the tendon of gluteus medius and the lateral aspect of the greater trochanter and the tendon of gluteus minimus. There is one, more remote from the joint, between the aponeurotic expansion of gluteus maximus over the tendon of the gluteus medius and greater trochanter, and the upper vastus lateralis.

    iii. **medially**, the bursa over the lesser trochanter behind the insertion of iliopsoas and in front of adductor magnus.

    iv. **posteriorly**, four. There is one between the piriformis and gluteus medius, one deep to the tendon of piriformis, one deep to the tendon of obturator internus as it emerges through the lesser sciatic foramen, and one, more remote, deep to gluteus maximus over the ischial tuberosity.

They can become inflamed and enlarged, in, for example, rheumatoid arthritis.

305. What muscles are attached to the greater trochanter ?

Gluteus medius is attached along a line passing downwards and forwards from the upper **posterior** part of the lateral surface of the trochanter. It is separated from the lateral surface of the trochanter in front of its insertion by the bursa of the gluteus medius.

Gluteus minimus is attached along a line passing downwards and laterally from the upper **anterior** part of the lateral surface of the trochanter, down to the lower insertion of gluteus medius. It is separated from the anterior edge of the trochanter by the bursa of gluteus minimus. Its insertion is separated from that of the gluteus medius by an area of bone covered by the bursa of the gluteus medius.

Piriformis is inserted principally into a facet on the posterosuperior, slightly inturned apex of the trochanter, and by a fibrous expansion passing anteriorly along the upper crest of the trochanter, lateral to, and then anterior to, the insertion of obturator internus.

Obturator internus and the **gemelli** are inserted into the medial slope of the trochanter, anterior to the main tendon of piriformis.

Obturator externus is inserted into the digital or trochanteric fossa, a depression in the medial part of the junction of neck and trochanter posterior to the site of insertion of obturator internus and the gemelli. The epiphyseal line of the trochanter passes through the floor of the fossa.

The upper fibres of **quadratus femoris** attach to the quadrate tubercle, which sits astride the epiphyseal line of the trochanter posteriorly.

306. What is inserted into the lesser trochanter ?

The psoas tendon is inserted into the medial aspect of the lesser trochanter. Iliacus muscle fibres are inserted into the anterior part of the lesser trochanter and into the bone below it, between the spiral and pectineal lines. The posterior aspect of the lesser trochanter is bare, covered only by a bursa between it and the upper fibres of adductor magnus. When the lesser trochanteric epiphysis is avulsed by strong action of the psoas against resistance, the avulsed fragment is prevented from excessive proximal displacement by the distal attachment of the muscular fibres of iliacus.

307. Where on the femur are the muscles attached that adduct the hip joint ?

Adductor longus has a linear insertion into the linea aspera in the middle third of the bone. Its insertion blends with vastus medialis anteriorly, and adductors brevis and magnus posteriorly.

Adductor brevis has a linear insertion from just below the lesser trochanter to the linea aspera, and along the proximal part of the linea aspera, posterior to pectineus and adductor longus and anterior to adductor magnus.

Adductor magnus has as linear insertion along the medial margin of the gluteal tuberosity, the linea aspera and the proximal part of the medial supracondylar line, and a rounded tendinous insertion into the adductor tubercle.

Pectineus has a linear insertion from the lesser trochanter to the linea aspera.

308. Where on the femur are the muscles attached that extend the hip joint ?

The deep part of the lower half of the **gluteus maximus** is inserted into the gluteal tuberosity, below the quadrate tubercle, between the insertions of adductor magnus and vastus lateralis. The gluteal tuberosity, so called, may be an elongated depression rather than a ridge, or may be prominent enough to be called the third trochanter. The adductor tubercle receives the adductor magnus from the ischial tuberosity.

309. From where on the femur does the quadriceps femoris muscle arise ?

Vastus lateralis has a linear origin from the upper end of the intertrochanteric line, down the anterior and inferior border of the greater trochanter, the lateral edge of the gluteal tuberosity, and the proximal half of the linea aspera.

Vastus medialis also has a linear origin from the lower end of the intertrochanteric line, the spiral line, the medial lip of the linea aspera, and the proximal part of the medial supracondylar line. It also arises further distally from the tendon of adductor magnus inserted into the adductor tubercle.

Vastus intermedius arises from the proximal two thirds of the anterior and lateral surfaces of the shaft.

310. Where on the femur are the muscles that flex the knee attached ?

The short head of **biceps femoris** has a linear origin extending down from just below the gluteal tuberosity. It arises from the lateral lip of the linea aspera, between the insertion of adductor magnus and the origin of vastus lateralis, and from the upper part of the lateral supracondylar line.

Gastrocnemius has two heads. The medial head arises from the back of the adductor tubercle and the part of the popliteal surface of the bone above the medial condyle. The lateral head arises from the lower part of the lateral supracondylar line and the facet of the bone below it.

Popliteus is attached by its round tendon to the anterior end of its groove, just below the lateral epicondyle.

Plantaris arises from the distal part of the lateral supracondylar line, above the origin of the lateral head of gastrocnemius.

311. What important arteries are closely applied to the femur ?

The **femoral artery**, as it passes through the subsartorial canal, approaches the shaft of the bone in the angle between the origin of vastus medialis and the insertions of adductor longus and magnus. Just before it passes through the hiatus of the adductor magnus, close to the bone, to become the popliteal artery, it gives off the descending genicular artery, which in turn gives muscular and articular branches close to the knee. The femoral artery becomes the **popliteal artery** which descends against the popliteal surface of the femur, and posterior aspect of the knee joint. It gives off medial and lateral superior genicular branches which encircle the sides of the supracondylar part of the bone to the anterior aspect of the knee joint.

The **profunda femoris artery** is the main branch of the femoral artery in the thigh. It passes posteriorly from the femoral triangle between pectineus and adductor longus. It runs down close to the linea aspera between adductor longus and adductor brevis above, and adductor longus and magnus below. Its **medial circumflex branch**, given off in the femoral triangle curves medially round the femur between pectineus and psoas major, and between adductor brevis and obturator externus. The **three perforating branches** of the profunda femoris pass backwards close to the linea aspera under tendinous arches of muscle to reach the vastus lateralis. The first usually passes above, the second through and the third below the insertion of adductor brevis, and then through adductor magnus and the origin of the short head of biceps. The profunda ends in adductor magnus as the fourth perforating artery.

The **nutrient artery** is quite large, arising usually from the second perforating artery. Alternatively two nutrient arteries may be present, one from the first and one from the third perforating arteries. They pierce the bone close to the linea aspera and are directed proximally.

The arteries close to the bone are subject to compression and injury from fractures, and from fracture-separation of the distal femoral epiphysis or dislocation of the knee joint.

312. How does the lower end of the femur develop ?

The single distal femoral epiphysis is unique in several ways. It is the first epiphysis to ossify and is present at birth. It is the largest and fastest growing epiphysis in the body. With the exception of the epiphysis of the medial end of the clavicle it is the last epiphysis to fuse with its diaphysis. Distal marginal irregularities of the epiphysis, seen on X-ray films, are common in the first decade, as might be expected in such a fast growing zone. They may appear only as a degree of roughening of the surface, or sometimes larger localised irregularities are apparent. Sometimes there is the appearance of an independent island of bone within an indentation suggestive of the condition of osteochondritis dissecans that is seen at a later age. These irregularities are not usually apparent in girls after about 5 years and boys after 10 years of age. When well formed, the growth plate has a slightly complicated configuration. The bony surface of the epiphysis presented to the growth plate has four cup-shaped depressions between anteroposterior and transverse ridges. Into the depressions fit four mamillary processes of the metaphysis separated by anteroposterior and transverse grooves. The growth plate is moulded between these surfaces. This arrangement contributes to the stability of the epiphysis to shear and torsional strains. It also possibly exposes the growth plate to more damage when separation does occur, so that growth disorder can be a subsequent complication. The growth plate passes through the adductor tubercle. Anteriorly it reaches the surface just above the articular surface. On each side it dips a little towards the condyles. The lower femoral growth plate contributes about two thirds of the length of the adult femur, and about one third of the length of the leg. The growth of the lower end of the bone is such that in the adult the line of the shaft of the femur is set at an angle of about 7° from the vertical line of the tibia, an angle used in the design of knee joint replacement prostheses.

313. Where is the capsule of the knee joint attached to the femur ?

What can be considered as the surgical capsule of the knee joint is now conceived as a three-layered complex of, from superficial to deep, deep fascia, ligaments and tendinous expansions, and an anatomical or true capsule. In some parts of the joint these layers can easily be distinguished, but in other parts they are fused together.

The superficial layer, or layer 1, consists of the fascia lata, continuous below with the fascia cruris. Posteriorly it overlies the bellies of gastrocnemius and the structures of the popliteal fossa. Medially it invests sartorius and fuses with its aponeurotic tendon and expansion into the periosteum of the tibia. Anterolaterally it receives an expansion of the iliotibial tract. Posterolaterally it fuses with the superficial part of biceps femoris and its expansion. Anteriorly it covers the extensor mechanism. It is not attached to the femur.

The intermediate layer, or layer 2, consists of —

i. the medial and lateral retinacular expansions of the quadriceps anteriorly.

ii. the anterior, vertical component of the medial collateral ligament, medially. The posterior oblique component of this ligament fuses with the deep layer posteromedially.

iii. what is usually described as the expansion of the insertion of the semimembranosus tendon, which, posteriorly, passes upwards and laterally, fusing with the deep layer to form the oblique popliteal ligament.

The deep layer, or layer 3, is the anatomical or true capsule. It is attached to the posterior articular margins of the femoral condyles and to the intercondylar line. Medially and laterally the capsule is attached a little way from the articular margins. It encloses the groove for and the tendinous insertion of the popliteus laterally. Anteriorly some would say the true capsule does not exist. Others describe the thin fibrous layer immediately superficial to the synovial membrane of the suprapatellar pouch as representative of the true capsule. It is attached to the anterior surface of the femur at the articular margin. On each side of the patella this membrane can be separated from the retinacular tendinous sheet as a rather soft layer, which splits to enclose the infrapatellar fat pad.

314. What ligaments stabilise the knee joint ?

In summary these are as follows. Medially is the medial collateral ligament. The anterior vertical fibres are separate from the true capsule. The posterior oblique fibres of this ligament fuse with the capsule and are sometimes distinguished as the separate posterior oblique ligament (not to be confused with the oblique popliteal ligament). Laterally is the lateral collateral ligament. Posterolaterally are the arcuate and deep lateral ligaments, sometimes described together as the posterolateral complex. When a fabella is present in the lateral head of gastrocnemius the deep lateral ligament is replaced by a fabellofibular ligament. Anterolaterally is the anterolateral femorotibial ligamentous part of the iliotibial band. Forming the central pivot of the joint are the anterior and posterior cruciate ligaments.

See questions 315, 332, 333 and 328.

315. Where are the ligaments of the knee joint attached to the femur ?

On the medial side, the proximal end of the medial collateral ligament (both its vertical and posterior oblique components) is attached to the localised area of the epicondyle.

On the lateral side the proximal end of the lateral collateral ligament is attached to the small area of the epicondyle, just above the insertion of the tendon of popliteus.

Posterolaterally strong capsular fibres pass from the epicondyle directly to the fibula as the deep lateral ligament. In 8–16% of knees a sesamoid, instead, called the **fabella,** is the centre of a fibrous complex of the true capsule deep to the lateral belly of gastrocnemius. It is attached to the femur through the posterior capsule proximal to it. The capsule distal to the fabella is thickened to form the **fabellofibular ligament,** attached to the head of the fibula.

Anterolaterally the deep fibres of the distal end of the iliotibial tract pass from the distal end of the attachment of the lateral intermuscular septum into the lateral supracondylar line, to Gerdy's tubercle, the smooth prominence on the anterior aspect of the lateral tibial condyle. These fibres contribute significantly to the anterolateral stability of the knee joint and have become known as the anterolateral femorotibial ligament.

Centrally, the anterior cruciate ligament is attached to an area on the far posterosuperior aspect of the intercondylar surface of the lateral femoral condyle. The posterior cruciate ligament is attached to a semicircular area in line with the articular margin on the anterior aspect of the intercondylar surface of the medial femoral condyle. The meniscofemoral ligaments vary as to presence and size. The posterior is the most constant, can be half the diameter of the posterior cruciate ligament, and is inserted into the posterior part of its attachment.

316. What muscles act as dynamic stabilisers of the ligaments of the knee joint?

Anteriorly the **quadriceps** mechanism is all important. Medially there is the tendinous complex, known as the pes anserinus, formed by the distal attachments of **sartorius, gracilis** and **semitendinosus.** Anteromedially the vastus medialis contributes to the medial patellar retinaculum. As well as its insertion into the tibia the **semimembranosus** tendon gives rise to extensions to the posteromedial aspect of the capsule, to the deep fascia of the lower leg and the posterior capsule. **Popliteus** tenses the posterolateral capsuloligamentous complex. The **biceps femoris** tendon inserts some fibres into the posterior part of the distal part of the lateral collateral ligament. The bellies of **gastrocnemius** tense the posterior capsule. The iliotibial tract is tensed to some degree by the **gluteus maximus** and **tensor fasciae latae**.

317. What is the nerve supply of the knee joint ?

Posteriorly there is a plexus of nerves in the fat superficial to the oblique popliteal ligament. It is formed from the fibres of two articular nerves. A branch from the **tibial nerve** arises above or within the popliteal fossa and passes downwards and laterally round the popliteal vein and artery to reach the plexus. The terminal branch of the **obturator nerve**, more commonly from its posterior division than its anterior, accompanies the femoral and popliteal arteries to the plexus. Branches from the plexus supply the capsule and pierce it to reach the peripheral parts of the menisci, the cruciate ligaments and infrapatellar fat pads.

Anteriorly there are **six** articular nerves. A branch from the **nerve to vastus medialis** divides deep to the medial patellar retinaculum to supply the anteromedial capsule. A branch from the **nerve to vastus lateralis** passes beneath it to the upper lateral capsule. Branches from the **nerves to vastus intermedius** reach the suprapatellar pouch. An upper branch from the **common peroneal nerve** above and behind the head of the fibula supplies the lower part of the lateral part of the capsule and the lateral collateral ligament. A lower branch curves upwards within the peroneus longus to enter the capsule at the anterolateral joint line. The infrapatellar branch of the **saphenous nerve** approaches the anterior aspect of the joint from the medial side just below the joint line, over, under or through the tendon of sartorius. It supplies the inferomedial part of the capsule and the patellar tendon. It also helps supply the skin over the front of the knee. Branches of the femoral and obturator nerves thus supply both hip and knee joints, so that disease in one joint can cause discomfort referred to the other. The infrapatellar branch of the saphenous nerve may be cut in an incision to open the medial side of the knee joint.

318. What are the features of the suprapatellar pouch ?

The suprapatellar pouch is derived from a true bursa which exists in the fetus between the lower end of the femur and the quadriceps tendon and muscle. The septum between the bursa and the synovial cavity of the knee breaks down in infancy, establishing the normal continuity.

The suprapatellar pouch extends about four finger-breadths above the upper pole of the patella. It covers the front, and the anterior thirds of the side surfaces, of the lower femur above the epiphyseal plate. The upward extent of the pouch can be appreciated when an effusion develops in the joint. Its size is significant for several reasons —

i. a wound of the skin over the area of the pouch may signify penetration into the synovial cavity, and is therefore a potentially serious injury.

ii. an infection in the lower femoral metaphysis may spread into the knee joint.

iii. the insertion of external fixation device pins through the pouch will lead to its obliteration. On the relatively infrequent occasion that a traction pin needs to be passed across the supracondylar part of the femur it should be done through the posterior half of the bone to give the best chance of avoiding the synovial cavity. At the same time the slight concavity of the popliteal surface of the bone, and the position of the popliteal vessels behind the knee should be kept in mind.

Seen through the arthroscope the suprapatellar pouch is a smooth-walled cavity. The attachments of **articularis genu** are seen as several patchy brownish areas at the upper limit of the cavity. A fold of synovium projects into the cavity from the medial side, extending backwards and downwards from the back of the quadriceps tendon to the front of the femur. It is the **medial suprapatellar plica** and may have a thick free edge. It may be more prominent and form a transverse membrane projecting inwards from the whole circumference of the pouch, leaving a small opening or porta between the two parts of the synovial cavity, or rarely it may form a thin, but complete septum, separating a suprapatellar bursa from the synovial cavity of the knee joint. The medial suprapatellar plica and the suprapatellar membrane are thought to be vestiges of the embryonic partition between the two cavities.

THE PATELLA

319. How can one orientate the patella ?

The patella is roughly triangular. The apex points downwards, and the curved base upwards. The anterior surface is convex and is longitudinally striated. Vascular foramina are present across the middle third of this surface. The posterior surface presents the articular area for the femur and is divided by a smooth vertical ridge into medial and lateral facets. The lateral facet is most often the larger. The back of the lower pole is bare of articular cartilage and has numerous vascular foramina.

320. What does the back of a fresh young adult patella look like ?

The articular cartilage is thick, particularly medially, where it is the thickest in the body. The articular surface does not reach the lower pole, which is covered posteriorly by synovium, fat, blood vessels and some tendon fibres. There is a smooth low vertical ridge dividing the articular surface into lateral and medial facets. Usually the lateral facet is the larger and is gently concave, to correspond to the more prominent lateral femoral condyle. Sometimes the medial and lateral facets are of equal extent, and sometimes the medial facet is very small or may even be absent. The relative sizes of the two facets serve as the basis for a classification of patellae into a number of different types. In addition, there is usually distinguishable a narrow area of articular cartilage on the most medial side of the surface, known to orthopaedic surgeons as the odd facet. This is the area of cartilage which most frequently displays the earliest signs of degeneration, which can often be seen even in young subjects at arthroscopy.

321. What does the patella do ?

i. The patella protects the femoral articular surface from direct trauma. When kneeling the weight of the body is borne on the tuberosity of the tibia, the patellar tendon and the lower part of the patella. These areas are additionally protected by the superficial infrapatellar and the prepatellar bursae.

ii. It lengthens the lever arm of action of the quadriceps femoris muscle from the axis of flexion and extension of the knee joint, so that the power of the muscle is enhanced and the power of extension of the tibia is maintained through to the last part of the arc of the movement.

iii. The main mass of quadriceps femoris pulls along the line of the shaft of the femur. This is at an angle of about 7–10°, known as the Q angle, to the vertical line of the tibia. The patella, tracking in the groove of the femur, serves to realign the pull of the quadriceps from the Q angle to the vertical.

322. How does the patella move in its track ?

In full extension of the knee, with the quadriceps relaxed, the central part of the articular surface of the patella rests gently on the front of the trochlear groove of the femur. When the quadriceps is tensed, however, only the lower part of the articular surface of the patella is in contact with the upper part of the groove. As flexion begins the articular surfaces make contact over a transverse band-like area. This area of contact migrates upwards on the patella and downwards and backwards on the femur, as flexion progresses. At 45° of flexion the contact band lies across the middle of the patella, occupying the lateral and medial facets, but not the odd facet. At 90° of flexion the contact band covers the upper part of the lateral and medial facets, but still not the odd facet. At 135° of flexion, however, the medial facet ceases to make contact. The patella rotates slightly medially on its long axis so that its anterior surface faces inwards. Contact is then borne by the upper two thirds or so of the lateral facet, and the whole of the odd facet. The contact areas on the femoral articular surface, in this final position of flexion, encroach onto the tibiofemoral surfaces, on facing areas on the anterior parts of the condyles. Also, as flexion progresses, the patella angles backwards in relation to the axis of the tibia. It is suggested that the deep infrapatellar bursa functions to assist the movement that results between the patellar tendon and the tibia above the tubercle, near the insertion of the patellar tendon.

323. What keeps the patella in its track ?

The stability of the patella in the trochlear groove of the femur depends on —
i. the **depth of the groove** between the condyles, in effect on the

prominence of the lateral condyle. Normally this is sufficient to present an obstacle to lateral subluxation of the patella beyond about 30° of flexion. A low lateral condyle can predispose to subluxation of the patella.

ii. the **size and shape** of the patella. The patella is normally of sufficient size, and the lateral facet of a contour, to correspond to that of the lateral condyle. A small patella is more likely to sublux than a larger one.

iii. the **position** of the patella. Normally the patella quickly settles into its groove on flexion of the relaxed, extended knee. If the patella is small, and higher than normal (patella alta), or if the knee is one which starts from a recurved position with lax ligaments, or if the patella is already a little displaced laterally as in knock-knee (genu valgum) deformity or persistent anteversion of the femoral neck, there is a predisposition to subluxation.

iv. the efficiency of the **vastus medialis**. The lower fibres of the muscle insert by short tendinous fibres into the superomedial border of the patella. They are sometimes described as forming a separate component of the muscle — the vastus medialis obliquus. Contraction of these fibres, especially in the last 30° or so of extension serve to counteract lateral displacement of the patella.

v. the checkrein action of the **retinacula**. Each retinaculum is a feltwork of vertically, obliquely and transversely directed fibres. Well defined oblique and transverse fibres constitute the patellofemoral ligaments which form palpable thickenings of the retinacula. On each side they pass in the retinaculum from the femoral epicondyle to the side of the patella. On the lateral side they are reinforced by an expansion from the iliotibial tract, which is called the iliopatellar band. On the medial side they are deep to the vastus medialis. Rarely the fibres on the lateral side can be too tight and then may sublux the patella on flexion of the knee. Other thickenings of the retinacula, called patellotibial ligaments, on either side of the patellar tendon, serve also to guide the patella in its groove.

THE TIBIA

324. Looking at just the upper end of the tibia, what features indicate the side of the body to which the bone belongs ?

The tuberosity faces forwards. The posterior aspect is indicated by the flat triangular area for the popliteus muscle, the backward overhang of the condyles and the 7–10° downward and backward slope of the weight bearing surface. The medial condyle is distinguished by the oval profile of its articular surface, and the horizontal groove for insertion of the semimembranosus tendon into its posterior part. The lateral condyle is distinguished by the circular profile of its femoral articular surface, the articular facet for the head of the fibula facing downwards and slightly backwards from below the projecting posterolateral part of the condyle, and the presence of the slight prominence that has become known as Gerdy's tubercle for the insertion of the iliotibial tract on its anterior aspect.

325. How does quadriceps femoris insert into the tibia ?

Quadriceps inserts into the tibia by intermediate attachment to the patella, which is a sesamoid bone within the tendon of the muscle. The rectus femoris forms a narrow flat, central and superficial tendon which mostly continues over the anterior surface of the patella into the patellar tendon. The vastus medialis becomes tendinous close to its insertion into the superomedial margin of the patella. The vastus lateralis becomes tendinous at a more proximal level before its insertion into the superolateral margin of the patella. The vastus intermedius forms a deep complex of tendinous fasciculi inserting into the upper margin of the patella. The composite of tendinous fibres proximal to the patella comprises the **quadriceps tendon**. The **patellar tendon** is formed mostly by the fibres of rectus femoris passing over the anterior surface of the patella. Sometimes fibres from the vastus lateralis contribute to the tendon. The mass of the tendon is inserted into the smooth area of the tibial tuberosity beginning just below its upper limit. Superficial fibres continue distally to cover the rougher lower part of the tuberosity. Medial tendinous fibres of the vastus medialis pass medial to the patella to contribute to the medial retinaculum, mixing with fibres of the fascia lata. Lateral fibres of the vastus lateralis pass lateral to the patella to contribute to the lateral retinaculum, mixing with fibres of the fascia lata and fibres from the iliotibial tract. The retinacula insert into the borders of the anterior surface of the tibia which slopes down from the front and sides of the condyles to converge on the tuberosity. Limited active extension of the knee is still possible after avulsion of the tibial tuberosity if the patellar retinacula remain intact.

326. What structures are attached to the plateau of the tibia ?

Structures are attached to both the periphery and the intercondylar area of the tibial plateau.

To the periphery of the plateau is attached the **true capsule,** or what can be called the deep layer of the three-layered surgical capsule, of the knee joint. Medially and laterally the attachments of the true capsule to the tibia are to the articular margins of the plateau. The short, reinforced portions of these parts of the capsule, between the attached peripheral borders of the menisci and the tibial plateau, are known as the coronary ligaments. Posterolaterally the lateral coronary ligament is interrupted by an hiatus for the tendon of the popliteus muscle. Posteriorly the true capsular attachment to the tibia passes across the posterior part of the posterior intercondylar area 2–3 mm below the rim, to include the most distal attachment of the posterior cruciate ligament within its boundary. Anteriorly the true capsule is more difficult to define, but its attachment is described as passing across the anterior part of the anterior intercondylar area. It is said to split around and thus enclose the anterior fat pad in a thin fibrous membrane, and is separated from the patellar tendon by the deep infrapatellar bursa.

To the intercondylar area of the plateau are attached the following structures, from anterior to posterior —

i. the **anterior horn of the medial meniscus**, to a small depression in the anterior and medial part of the anterior intercondylar area.

ii. the **anterior cruciate ligament**, to a smoother area, behind the attachment of the anterior horn of the medial meniscus, anterior to the intercondylar eminence.

iii. the **anterior horn of the lateral meniscus**, to lateral to the attachment of the anterior cruciate ligament, just anterior to the intercondylar eminence.

iv. the **posterior horn of the lateral meniscus**, to the posterior slope of the intercondylar eminence.

v. the **posterior horn of the medial meniscus**, to a depression in the posterior intercondylar area behind the medial intercondylar tubercle, posteromedial to the attachment of the posterior horn of the lateral meniscus.

vi. the **posterior cruciate ligament**, to the major remaining part of the posterior intercondylar area, back to the insertion of the capsule into the tibial plateau, just below its rim.

327. Looking into the front of an open knee joint, what signs would indicate to which side of the body it belongs ?

The lateral femoral condyle is the more prominent. The anterior cruciate ligament passes backwards, upwards and laterally. The lateral meniscus is a larger arc of a smaller circle. The medial meniscus is a smaller arc of a larger circle and is characteristically shaped, being wider behind than in front.

328. What do the cruciate ligaments do ?

The cruciate ligaments are essential to the stability and normal movement of the knee joint.

i. They provide anteroposterior stability. The anterior cruciate ligament prevents backward displacement of the femur on the tibia or forward displacement of the tibia on the femur. The posterior cruciate ligament acts in a complementary way.

ii. They resist medial rotation of the tibia on the femur, becoming 'wound up' in this movement.

iii. They resist hyperextension.

iv. They are essential to the normal rhythm of movement of the articular surfaces in flexion and extension. There are only three ways that a curved articular surface can move on a flat surface. It can spin on its mechanical axis, the position of the point of contact remaining stationary. Or it can slide, its point of contact translating in the plane of the joint. Or it can roll, its points of contact describing an arc on its surface. In flexion of the knee joint only slide and roll can occur. If only roll occurred the femoral condyles would fall off the back of the plateau of the tibia beyond about 60° of flexion. If only slide occurred the posterior rim of the tibial plateau would be halted before full flexion occurred by impingement against the popliteal surface of the femur. The ratio of slide to roll of the articular surfaces, to allow the normal range of movement from about 5° of hyperextension to 150° of flexion, is determined by the exact geometry of the cruciate ligaments, in their lengths and sites of attachment, providing what mechanical engineers call a four-bar linkage system. The ratio of slide to roll increases from about 2:1 to about 4:1 as flexion progresses.

v. They help guide the articular surfaces into the locked or close-packed position, of maximum stability, of the joint, during the last 15° of extension, when the femur medially rotates on the tibia.

vi. The posterior cruciate ligament, the slightly shorter but stronger and more tear-resistant of the two, provides the pivot of rotation in the latter part of flexion.

329. What is the structure of a meniscus ?

Each meniscus is a crescent of fibrocartilage. In radial section each has the shape of a curved wedge. It is attached by its horns to the intercondylar area, and peripherally to the capsule of the joint. It forms, with the intercondylar eminence, an osseo-meniscal ring, which helps support the load transmitted through a femoral condyle. The fibrils, of type 1 collagen, are arranged in fibres rather like tendon. Those forming the bulk of the meniscus are arranged circumferentially around the crescent. They resist the tension – known as hoop stress – that develops in the meniscus when the weight of the femoral condyle tends to increase the radius of the curvature of the crescent. The fibres at the surfaces of the meniscus, to a depth of about 100 microns, are arranged principally in radial fashion. They may be designed to prevent the circumferentially arranged fibres from separating, and thereby causing a split in the meniscus. Each meniscus is supplied by radial branches from a perimeniscal capillary plexus derived from branches of the upper and lower genicular arteries. The radial vessels penetrate the stroma of the periphery of the meniscus to a depth of about 10–25% of its width. This blood supply enables a tear of the periphery of a meniscus to effect repair, a process which is not possible in the inner avascular zone. Vessels from the middle genicular artery enter the attachments of the horns of the menisci to the intercondylar area via vascular synovium. Vessels within a meniscus end in capillary loops. There is an avascular area of the periphery of the lateral meniscus next to the tendon of the popliteus muscle.

330. What do the menisci do ?

i. **The menisci of the knee play an important role in load transmission**. Their horns are secured near the median line of the tibial plateau. Two resilient osseo-meniscal rings are thus formed, which receive the condyles of the femur. The lateral meniscus forms a greater part of the lateral ring than does the medial meniscus of the medial ring. The medial tubercle of the intercondylar eminence forms a more prominent part of the medial ring than does the lateral tubercle of the lateral ring. The area of tibial articular cartilage exposed within the inner margin of the medial meniscus is about twice that exposed within the lateral meniscus. It has been shown experimentally that, under conditions of no loading (i.e. at rest, non-weight bearing), the menisci slightly separate the opposing central cartilaginous articular surfaces from each other, and that most contact of the femoral condyles is made at the periphery of the menisci. Only as load increases do the central cartilaginous surfaces come into significant contact. The areas of contact and load bearing spread first to the cartilage covering the medial intercondylar tubercle, and to the inner aspects of the menisci. Only then are the central areas exposed to the full compressive force of the femoral condyles. The force tends to dilate the apertures of the rings. It is resisted by the tension of the circumferentially arranged collagen fibres of the main masses of the menisci, thereby protecting the cartilaginous articular surfaces from increasing force over a small area. It is thought that menisci transmit in this way between 60–70% of the load through the knee joint. The importance of this function of a meniscus is borne out clinically by the development of degenerative changes in a joint within a few years of excision of a complete meniscus following an injury. A break in the ring of a meniscus also renders it relatively functionless in load bearing and can lead to the same result.

ii. **They probably enhance the stability of the joint**. In marked rotation there is a tendency for a condyle to ride out of its meniscal ring. The passive distraction of the articular surfaces that results is resisted by ligaments, and by the reflex contraction of the muscles which give dynamic support to the joint.

iii. They are thought to **assist in lubrication** of the articular surfaces with synovial fluid, and by acting as space-fillers to prevent entrapment of synovium and capsule.

331. How do the menisci move within the knee joint ?

When the knee flexes, the contact areas between the femur and tibia move backwards, and in so doing displace the meniscal rings, still attached at their horns, in the same direction. The meniscal rings become slightly distorted, especially the lateral one. Instead of outlining round areas of articular cartilage on the tibia, the menisci assume a rather ovoid shape, the medial meniscus directed posteromedially and the lateral meniscus directed posterolaterally.

An arc of some 50° of rotation of the tibia on the femur in the flexed position is often possible, though there is variation between subjects. The menisci move with the femoral condyles on the tibia. The axis of rotation is to the posteromedial side of the centre of the tibial plateau, and as flexion progresses it approaches the line of the posterior cruciate ligament, which then becomes the pivot of the joint. The lateral meniscus thus moves to a greater extent than the medial meniscus. Additional gliding surfaces for the menisci on the tibia are provided for these movements anteriorly and posteriorly. They are relatively small for the medial meniscus, but are more extensive for the lateral meniscus.

The lateral meniscus is more mobile than the medial meniscus. Additional control of the lateral meniscus is necessary if its entrapment between the joint surfaces is to be avoided when the knee flexes. The medial part of popliteus is inserted into the back of the meniscus. When it contracts it pulls the meniscus out of the path of the advancing lateral femoral condyle. This is especially important in the first part of flexion, when the joint unlocks from a position of some medial rotation of the femur on the tibia. The meniscofemoral ligaments also help to guide and restrain the posterior part of the lateral meniscus in its posterior displacement. The most constant of these arises from the posterior part of the periphery of the meniscus. It passes parallel to and behind the posterior cruciate ligament to the lateral aspect of the medial surface of the medial femoral condyle. It is also called the ligament of Wrisberg and has even been designated the third cruciate ligament. A less constant and thinner band passes in front of the posterior cruciate ligament and is known as the ligament of Humphrey. It is possible that the anterior transverse menisco-meniscal ligament often plays a role also in restraining the displacement of the anterior horns of the menisci during rotation. It passes between the anterior margins of the menisci and is a well developed structure in about 40% of knees. It is absent, however, in about another 40% of knees. A posterior menisco-meniscal ligament, passing in front of the posterior cruciate ligament, is present in only about 2% of knees.

The lesser mobility of the medial meniscus, together with the efficient control of the lateral meniscus, probably explains the greater vulnerability to tear shown by the medial meniscus. The medial meniscus is up to ten times more likely to tear than the lateral meniscus.

332. What are the structures that stabilise the medial side of the knee joint, and where are they attached to the tibia ?

The stability of a joint depends on the extent, curvature and congruity of its articular surfaces and on the soft tissue supports around it. The knee joint offers no bony mechanism for stability and has to rely on soft tissue supports only. These are several —

i. there are capsular thickenings distinguished as ligaments.

ii. there are intracapsular and extracapsular ligaments, separate to varying degrees from the capsule. These are static structures dependent on their inert strength and geometry for their effectiveness.

iii. some ligaments are reinforced by the insertion of tendon or muscle. These insertions provide a dynamic component to the function of the ligaments. Such ligaments are accordingly described as 'dynamised'.

iv. there are tendons passing near to and inserting close to the joint.

On the medial side of the knee these separate elements comprise —

i. the **deep medial ligament**, a thickening of the inner aspect of the medial part of the capsule, inserting into the rim of the plateau of the tibia. The lower part of this ligament, between the meniscus and the plateau, is part of the medial coronary ligament.

ii. the **medial collateral (superficial medial) ligament**, consisting of two parts, of parallel and oblique fibres. The parallel fibres form a vertical band from the medial epicondyle of the femur, to an area on the tibia measuring about 1 cm wide by 5 cm long, on and in front of the medial border of the tibia, above the point at which the soleal line meets this border. It is separated from the underlying capsule and tibial condyle by one or more bursae. The oblique fibres pass backwards and downwards from the upper attachment of the parallel fibres in a fan-like manner to strengthen the capsular insertion into the posteromedial margin of the plateau. This part of the ligament is sometimes described as a separate entity, the **medial oblique ligament**. It is dynamised by fibres derived from semimembranosus, which can therefore also be considered to exert some control over the posterior part of the medial meniscus. It is also dynamised by some fibres of origin of the medial head of gastrocnemius. Some oblique fibres also pass upwards and backwards from the lower end of the parallel fibres, to the capsule at its insertion into the tibial plateau.

iii. the tendon of **semimembranosus**, which has an elaborate and widespread insertion. The bulk of the tendon is inserted into the back of the medial tibial condyle, to the medial end of the horizontal groove. Expansions of the tendon pass upwards to strengthen the medial capsule, anteriorly along the horizontal groove under the medial collateral ligament, downwards to the medial border of the tibia, downwards and laterally, to the soleal line, forming the fascial covering of the popliteus muscle, and upwards and laterally to strengthen the capsule as the oblique popliteal ligament.

iv. the tendons of **sartorius, gracilis**, and **semitendinosus** which pass superficial to the medial collateral ligament, before inserting into the upper end of the medial surface of the tibia. The sartorius tendon forms an expansion in the form of an inverted J which overlies the insertions of the tendons of gracilis and semitendinosus. It fuses with the overlying deep fascia and the underlying periosteum, and sometimes with the tendons of gracilis and semitendinosus, forming between them the structure known as the **pes anserinus**. The tendon of insertion of gracilis is above and in front of that of semitendinosus.

333. What are the structures that stabilise the lateral side of the knee joint and where are they attached to the tibia ?

The structures that stabilise the lateral side of the knee comprise the iliotibial tract, the lateral collateral ligament, the deep lateral or fabellofibular ligament, the arcuate ligament, the biceps femoris and popliteus.

Of these only the iliotibial tract, biceps femoris and popliteus attach to the tibia.

i. The main insertion of the **iliotibial tract** is into the prominence on the anterior aspect of the lateral tibial condyle, known as Gerdy's tubercle. Expansions of the tract pass forwards to reinforce the lateral patellar retinaculum, and distally to merge with the aponeurotic deep fascia over the upper part of tibialis anterior. The deep distal part of the iliotibial tract extends from the distal part of the lateral intermuscular septum attached to the proximal part of the lateral femoral condyle, and constitutes a femorotibial collateral ligament now being recognised as an important stabilising mechanism of the anterolateral part of the knee joint. The iliotibial tract is dynamised by the insertions into it proximally of tensor fasciae latae and gluteus maximus, distally of the tibialis anterior via its deep fascial covering, and posteriorly, especially in flexion, of an expansion from the biceps tendon.

ii. The **biceps tendon** inserts distally in a rather complex manner, by splitting into three laminae. The superficial lamina extends forwards to merge with the iliotibial tract, and spreads distally over the head of the fibula into the deep fascia of the leg over the extensor, peroneal and crural compartments. The extension forwards over the extensor compartment is particularly strong and reaches to the anterior border of the tibia, up to Gerdy's tubercle. The middle lamina ensheaths the lateral collateral ligament on both its deep and superficial surfaces as it inserts into the head of the fibula. It has a fibrous attachment to the posterior aspect of the ligament, helping to keep the ligament firm as it is relaxed on flexion of the knee. The deep lamina of the biceps tendon spreads out to insert into the posterior part of the lateral tibial condyle rather like semimembranosus, and into the side of the condyle around to Gerdy's tubercle. Fibres mingling with the capsule serve to reinforce it.

iii. **Popliteus** is attached distally to the medial two thirds of the triangular area on the upper part of the posterior surface of the tibia outlined by the medial border and soleal line. It is attached medially by tendinous fibres into the posterior border of the lateral meniscus, and laterally by a narrow tendon which passes through the capsule under the arcuate ligament and between capsule and synovium to the anterior end of the popliteal groove on the lateral condyle of the femur. The **arcuate ligament** comprises a thickening of the capsule, by fibres which pass medially from the head of the fibula, to mingle with the musculotendinous junction of the popliteus at the capsule of the joint.

334. What bursae are there around the knee joint in contact with the tibia ?

i. The superficial infrapatellar bursa overlies the patellar tendon and tuberosity. Its enlargement is recognised as clergyman's knee, distinguished from housemaid's knee, which is an enlargement of the prepatellar bursa higher up.

ii. The deep infrapatellar bursa intervenes between the deep surface of the patellar tendon and the tibial surface deep to it.

iii. There is a bursa between the tendons of sartorius, gracilis and semi-tendinosus, and the underlying medial collateral ligament and tibia.

iv. There is a bursa between the anterior vertical part of the medial collateral ligament and the underlying joint capsule and expansion of semi-membranosus. It may be above, across or below the joint line.

v. There is a bursa between the distal part of the tendon of semimem-branosus and the capsule and medial tibial condyle.

vi. The popliteal bursa lies between the upper part of the muscle and the underlying bone. It is an outpouching of synovium which passes down deep to the muscle from the hiatus for its tendon in the lateral coronary ligament.

335. How does the upper end of the tibia develop ?

At birth the upper end of the tibia is usually wholly cartilaginous, the analogue sloping anteriorly into a well formed tuberosity. A primary centre of ossification for the main part of the proximal tibial epiphysis may be present at birth, but it usually appears within the first couple of months. The epiphysis that develops is a very secure one, not prone to traumatic separa-tion from the metaphysis.

The factors that contribute to this stability of the proximal epiphysis are —

i. its slightly irregular contour. It is greater in height laterally than medially, and has a small central prominence fitting into a reciprocal depression in the metaphysis.

ii. the support on its lateral side by the buttress of the head of the fibula, articulating with the upper end of the tibia at the superior tibiofibular joint, above the line of the growth plate.

iii. the support of tendon and ligament. Posteriorly, the epiphysis lies above the popliteus. Medially, the growth plate is straddled by the insertion of semimembranosus, and the distal attachment of the medial collateral ligament is beyond it. Anteriorly, the superficial fibres of the patellar tendon and the patellar retinacula pass down beyond the growth plate to reach the periosteum of the metaphysis.

The development of the tuberosity is relevant to its susceptibility at different ages to traction injury. In the young child the fibres of the patellar tendon insert into the cartilaginous tuberosity and spread out radially to form fibrocartilage. One or more centres of ossification appear in the distal part of this fibrocartilage between about 10 and 13 years of age, earlier in girls than in boys. They form an apophysis, or traction epiphysis. During this stage the weakest part of the extensor mechanism is the bony nucleus of the secondary centre. Traction may cause it to fracture. The resulting local pain, tenderness, swelling and new bone formation presents as the clinical condition known as Osgood–Schlatter's disease. A year or so later a tongue of the proximal epiphysis grows down to fuse with the apophysis. The distal part of the deep attachment of the tongue of ossifying tuberosity to the metaphysis then gradually changes from above downwards from fibrocartilage to the columnar hyaline cartilage of the epiphysis of the proximal end of the bone. The apophysis has become part of the epiphysis. At this stage the weakest part of the extensor mechanism is the hypertrophic and calcifying zone of the epiphyseal plate. The epiphysis of the tuberosity may then avulse from the metaphysis through this layer.

The upper tibial epiphysis is second in size only to the lower femoral epiphysis. It contributes 55% of the length of the tibia and 27% of the length of the leg. It is said to grow at the rate of about just over 0.5 cm (a quarter of an inch) a year. The growth plate closes between about 16–18 years of age.

336. Where on the tibia are the muscles attached that dorsiflex the foot ?

Tibialis anterior arises from the upper half to two thirds of the lateral surface of the shaft of the bone, from the anterior border in front to the interosseous border behind, extending up to the lateral surface of the lateral condyle, between the sharp anterolateral border of the condyle and the interosseous border, to the margin of the superior tibiofibular joint.

A small part of **extensor digitorum** arises from the lateral condyle just above and in front of the superior tibiofibular joint.

337. Where on the tibia are the muscles attached that plantar flex the foot ?

Soleus has an aponeurotic origin from the middle third of the medial border of the bone, and upwards along the soleal line to where a tubercle, short of the interosseous border, marks the attachment of the fibrous arch of the muscle, which extends to its origin from the fibula.

Tibialis posterior arises from the posterior surface of the bone. Its upper limit of origin is demarcated by the interosseous border, the soleal line, and the vertical line. Its lower limit extends down to the junction of the middle and lower thirds of the shaft.

Flexor digitorum longus arises from the posterior surface of the bone from the area outlined above by the vertical line, the soleal line and the medial border, down to about the junction of the third and lower quarters.

Some fibres of **peroneus longus** arise from the lateral condyle above and behind those of extensor digitorum longus.

338. What arteries are close to the tibia ?

The **popliteal artery** passes down behind the upper part of the tibia separated from it by the popliteal fascia and popliteus, to the lower border of the muscle.

The **inferior medial genicular artery** is a branch of the popliteal artery which passes along the upper border of popliteus around the medial tibial condyle deep to the medial head of gastrocnemius and the medial collateral ligament, to reach the infrapatellar fat pad and patellar tendon.

The **inferior lateral genicular artery** is a branch of the popliteal artery which passes across popliteus around the lateral tibial condyle under the lateral head of gastrocnemius. It passes over the arcuate and fabellofibular ligaments, deep to the lateral collateral ligament of the knee, to the infrapatellar fat pad and patellar tendon. The inferior genicular arteries are vulnerable to injury in operations in this area.

The **anterior tibial artery,** in 90% of cases, arises by division of the popliteal artery at the lower edge of popliteus. It passes forwards between the upper ends of the tibia and fibula above the upper edge of the interosseous membrane. The popliteal artery and its branches are here vulnerable to distortion, for instance following a fracture separation of the upper tibial epiphysis, or after osteotomy of the upper end of the tibia to correct varus or valgus deformity. In the lower third of the leg the artery approaches the tibia, is crossed by the tendon of extensor hallucis longus, and passes under the superior extensor retinaculum between the tendons of extensor hallucis longus and extensor digitorum longus, midway between the malleoli. The artery is vulnerable in its distal course across the bone during the insertion of external fixation pins. In lateral profile an area of bone over a distance of about 4 cm beyond the junction of the third and lowest quarters of the tibia should be avoided in this procedure, if possible.

The **posterior tibial artery** crosses the lower end of the posterior surface of the tibia beyond its course behind the tibialis posterior muscle. It lies between the tendons of flexor digitorum longus and flexor hallucis longus.

The **nutrient artery** to the tibia arises from the posterior tibial artery and enters the bone on or close to the vertical line marking the boundary between the origins of tibialis posterior and flexor digitorum longus. It is directed distally to the marrow cavity.

339. How are the tendons arranged around the lower end of the tibia ?

In front of the lower end of the tibia the extensor tendons pass deep to the superior retinaculum and a compartment of the inferior extensor retinaculum. The **superior extensor retinaculum** is a thickening of the anterior deep fascia of the lower part of the leg, extending between the lower end of the anterior border and medial malleolus of the tibia, and the lower end of the anterior border and lateral malleolus of the fibula. Its upper and lower borders are rather indefinite. The oblique superomedial band of the **inferior extensor retinaculum** extends across the front of the ankle joint to the anterior border of the tibia and the medial malleolus.

From medial to lateral lie the tendons of tibialis anterior, extensor hallucis longus, extensor digitorum longus and peroneus tertius. The tendon of tibialis anterior often passes through a tunnel of the superior retinaculum, caused by a split of the retinaculum into deep and superficial layers. Distal to the extensor retinaculum the superomedial band of the inferior extensor retinaculum passes behind the tendon of tibialis anterior, but may also provide a superficial layer to it. The tendon of extensor hallucis longus passes deep to the superior and inferior retinacula. A deep layer of the inferior retinaculum forms a recurved loop around the tendon to insert into itself or the neck of the talus. The tendons of extensor digitorum longus and peroneus tertius pass deep to the superior extensor retinaculum in front of the lower end of the tibia.

Behind the lower part of the tibia the tendon of tibialis posterior passes across the bone to a groove on the posterior surface of the medial malleolus converted to a tunnel by a fibrous sheath. Immediately lateral to this is the tendon of flexor digitorum longus, which passes down posterior (i.e. superficial) to the tendon of tibialis posterior, and reaches its own fibrous sheath beginning behind the distal tibia. The tendon of flexor hallucis longus receives muscle fibres into its lateral side down to the lower limit of the tibia where it also crosses the ankle joint in a fibrous tunnel. The deep transverse fascia of the leg here stretches from the lower medial border of the shaft and posterior border of the medial malleolus to the posterior border of the lower fibula, overlying the deep tendons.

340. How does the lower end of the tibia develop ?

The secondary centre of ossification for the lower tibia develops as a horizontally aligned oval nucleus during the latter half of the first year of life. It lies at first closer to the medial side than the lateral side. From the age of about 4 years contouring of the growth plate begins. An upward projection develops from the anteromedial part of the epiphysis. Ridges also develop from the other corners of the epiphysis, all of which fit into corresponding depressions in the metaphysis and serve to secure the epiphysis against traumatic separation. Extension of the ossific nucleus into the medial malleolus begins at about 7 years of age and is not complete until the age of about 15. An accessory centre of ossification can appear in the cartilage of the medial malleolus at about the same time as bone begins to grow down from the main centre. Though the two centres usually fuse, the malleolar centre may remain as a separate ossicle, the os subtibiale, which should not be confused with a fracture. Closure of the lower tibial growth plate occurs between about the ages of 16 and 18 years. The lateral part of the plate is the last to close. The growth plate is entirely proximal to the capsule of the ankle joint. It contributes 45% of the length of the tibia and 18% of the length of the lower limb.

THE FIBULA

341. How can one orientate the surfaces of the fibula ?

The upper end, or head, has a rounded facet for articulation with the lateral tibial condyle. It faces upwards, and usually forwards and medially.

Posterolateral to the facet is the prominent styloid process. The lower end or lateral malleolus has a triangular facet for articulation with the talus. It faces medially, and slightly downwards. Behind the facet is a deep groove, sloping downwards and forwards, called the digital or malleolar fossa. The anterior border is found by continuing upwards the ridge formed at the apex of the triangular subcutaneous area of the lateral malleolus and lower shaft to the neck of the bone. The interosseous border is found, similarly, by continuing upwards the ridge formed at the apex of the triangular area above the facet for the tibia. It approaches the anterior border closely in the upper half of the bone. The posterior border is found by following upwards the posterior border of the digital fossa to the neck of the bone. It may be roughened in its upper part by the origin of soleus. The medial crest begins above near the upper end of the interosseous border, curves down prominently towards the posterior border, but before reaching it curves back to the interosseous border to meet it near the junction of the upper two thirds with the lower third of the shaft. Another vertical ridge may be present between the interosseous border and the medial crest. The extensor surface of the bone lies between the anterior and interosseous borders. The peroneal surface is between the anterior and posterior borders. The flexor surface lies between the posterior and interosseous borders.

342. How is the fibula attached to the tibia ?

The fibula is attached to the tibia by —

a) the **superior tibiofibular joint**. The articular surfaces are nearly flat. In 25% of cases the plane of the joint is almost horizontal; in the remainder the plane is variably oblique, and in some nearly vertical. Horizontal articular surfaces allow a little rotation of the fibula, caused during dorsiflexion and plantar flexion of the ankle by a slight anteroposterior convexity of the lateral talar articular surface. More vertically orientated superior tibiofibular joints are said to be associated with flatter lateral talar articular surfaces. Slight translation movements also occur at the superior tibiofibular joint. The capsule is reinforced anteriorly and posteriorly by bands of fibres running obliquely across the joint. The synovial cavity may communicate with that of the knee joint by an opening into the synovial recess deep to popliteus.

b) the **interosseous membrane** attached to the interosseous borders up to just below the neck of the fibula. The anterior tibial vessels cross above the membrane. Its fibres for the most part are short and directed downwards and laterally. Some fibres in its upper part pass downwards and medially, while others are vertically disposed behind the lower part of the membrane. It is pierced in its lower part by the perforating branch of the peroneal artery. It is continuous below with the interosseous ligament of the inferior tibiofibular joint. From its anterior surface partly arise tibialis anterior, extensor hallucis longus, extensor digitorum and peroneus tertius, while posteriorly it gives attachment to tibialis posterior and flexor hallucis longus. Anteriorly also lie the anterior tibial vessels and the deep peroneal nerve. Apart from providing a surface for muscular attachment, the membrane probably serves to prevent the outward bowing of the fibula which is caused by the pull of the muscles arising from it. Strain measurements of the tibia in cadaver experiments, with and without the membrane, show that the membrane does not transmit the stress of load bearing from the tibia to the fibula.

c) the **inferior tibiofibular joint**, an important part of the ankle complex. Its component parts are —

i. the **anterior, inferior tibiofibular ligament** — a flat band passing from the lower end of the anterior border of the shaft and the upper anterior border of the lateral malleolus of the fibula, upwards and medially to the anterolateral tubercle of the tibia. It is the first ligament to suffer the strain, and to rupture, in the common injury of the ankle, in which the talus rotates laterally in its mortise. It is the weakest of the three ligaments of this joint.

ii. the **posterior inferior tibiofibular ligament** — a strong low triangular band passing upwards and medially from the upper part of the digital fossa, from the tubercle above it, and the border of bone behind it. The superficial part of the ligament passes to the posterolateral tubercle of the distal end of the tibia. The deep part, also called the **inferior transverse ligament,** extends along the posterior border of the inferior articular surface of the tibia, as far as the medial malleolus. It forms a posterior labrum for the ankle joint, imprinting the posterior part of the lateral border of the talar articular surface. It is strong enough to pull off a 'third malleolus' from the tibia in lateral rotation injuries of the ankle. The terminal branch of the peroneal artery runs down across it.

iii. the **interosseous ligament**. It consists of short, strong fibres intermingled with fat and blood vessels, uniting the fibular groove of the tibia and the facing convex surface of the fibula. It forms the roof of a synovial recess which is present above small, facing, crescentic articular surfaces of the lateral malleolus and tibia, continuous with the articular surfaces and synovial cavity of the ankle joint.

343. What is the relationship of the head of the fibula to the knee joint ?

In the embryo the upper end of the fibula and tibia are at the same level, and are surrounded by the tubular capsule which encloses also the lower end of the femur. With differential growth the head of the fibula becomes excluded from the knee joint, sinking below the tibial plateau, and pulling down a prolongation of the original capsule, which in the mature knee forms the posterolateral ligamentous complex of the joint. Attached distally to the styloid process of the head of the fibula, this complex can be regarded as formed of fibres passing upwards in two directions. Arching medially, they form the **arcuate ligament**, which fuses with the medial part of the musculotendinous junction of the popliteus, which in turn is inserted into the posterior part of the lateral meniscus. Superiorly, they pass to the upper capsule and lateral epicondyle of the femur as the **deep lateral ligament**. If a fabella is present they pass to it to form the **fabellofibular ligament**. The lateral collateral (superficial lateral) ligament stands away from the deep part of the capsule. It is thought possibly to be the evolutionarily degenerate tendon of origin of peroneus longus. The head of the fibula can be excised without causing marked instability of the knee joint. The lateral collateral ligament is lax in flexion, and in extension the tension of the other ligaments, and the posterior capsule, are sufficient to provide stability.

344. What muscles attach to the extensor surface of the fibula?

i. **Extensor digitorum longus** and **peroneus tertius** have a continuous origin, from the head to about 2.5 cm above the lateral malleolus. They occupy the whole thin width of the surface except for the origin of extensor hallucis longus.

ii. **Extensor hallucis longus** arises from the middle two quarters of the length of the surface, medial to extensor digitorum longus and peroneus tertius.

345. What muscles attach to the peroneal surface of the fibula ?

i. **Peroneus longus** arises from the back of the head, the whole of the peroneal surface of the upper third of the bone and the posterior part of its middle third.

ii. **Peroneus brevis** arises from the anterior part of the middle third of the peroneal surface, and the whole of the lower third down to within 2–3 cm of the lateral malleolus.

The origins of the two peronei overlap therefore in the middle third of the leg.

346. What muscles attach to the flexor surface of the fibula ?

i. **Soleus** arises from the back of the head and neck, and the upper third of the flexor surface. Its lower origin is marked by a ridge of varying extent.

ii. **Tibialis posterior** arises from the area marked out between the interosseous border and medial crest. The vertical ridge between these boundaries gives rise to an intramuscular tendon.

iii. **Flexor hallucis longus** arises from the lateral half of the surface of the middle third, bounded by the origins of tibialis posterior and soleus, and from the whole width of the lower third, down nearly to the lateral malleolus.

347. What nerves are related to the fibula ?

The **common peroneal nerve** passes down behind and below the head of the fibula on the upper fibres of soleus, then around the neck of the bone, and through the upper origin of peroneus longus. In this exposed part of its course it is very liable to compression injury. It divides into deep and superficial peroneal nerves within peroneus longus. The **deep peroneal nerve** continues through extensor digitorum longus to the extensor compartment. The **superficial peroneal nerve** passes downwards at first close to the bone between peroneus longus and extensor digitorum longus, before piercing the deep fascia between them, at roughly the junction between the middle and lower thirds of the leg. The superficial peroneal nerve can be entrapped as it pierces the deep fascia.

348. What dynamic functions does the fibula perform ?

i. It transmits some of the weight passing through the lower leg. On theoretical grounds, because the tibiofibular surfaces are oblique to the vertical, some force must pass through the joints perpendicular to their surfaces. Strain gauge experiments on cadaver limbs suggest that about one sixth of the body weight passing through the leg is borne by the fibula.

ii. At the ankle it provides a strong but slightly malleable grip on the talus. Flexion and extension movements of the talus within the ankle mortise entail slight rotation of the fibula (2.5–12°) on its long axis at the superior tibiofibular joint, and some tilting of the malleolus as viewed from the front, with slight translation of the articular surfaces of the superior tibiofibular joint.

iii. By allowing slight movement of the fibula at both joints, distally brought about by contraction of the long muscles arising from the bone, and proximally by the insertion of biceps femoris into its head, some protection against fatigue fracture of the bone is provided.

349. Where are the septa between the compartments of the leg attached ?

The **anterior peroneal septum**, which separates the extensor and peroneal compartments, is attached to the anterior border of.the bone, between the extensor and peroneal surfaces. The **posterior peroneal septum**, which separates the superficial flexor and peroneal compartments, is attached to the posterior border between the flexor and peroneal surfaces. The **deep transverse septum** (or fascia) is also attached to the posterior border. It passes medially between soleus and the deep flexors to the soleal line and medial border of the tibia, dividing the posterior compartment into deep and superficial parts. The **interosseous membrane** attaches to the interosseous border. The extensor and peroneal compartments can be decompressed by an anterolateral approach, over the anterior peroneal septum. The deep and superficial posterior compartments can be decompressed by a medial approach behind the medial border of the tibia.

THE ANKLE AND FOOT

350. What are the features of the ankle mortise ?

The term mortise has come to be applied to the socket formed by the lower ends of the tibia and fibula into which the talus fits like a tenon. In detail, the mortise comprises —

i. the vertical, flat, comma-shaped, lateral articular surface of the medial malleolus.

ii. the horizontal, lower, weight-bearing articular surface of the tibia, described as the **plafond** or vault of the ankle joint. It is concave anteroposteriorly and slightly convex transversely. There is a slight antero-posterior elevation dividing the surface into a wider lateral and a narrower inner segment. Slight anterior and posterior lips limit the surface. The posterior lip is accentuated by the labrum-like attachment of the deep part of the posterior inferior tibiofibular ligament.

iii. the vertical, triangular, medial articular surface of the lateral malleolus, slightly convex vertically and concave horizontally.

iv. a linear opening leading into a small recess of the inferior tibiofibular joint between the lateral malleolar facet and the plafond of the tibia. Narrow crescentic strips of articular cartilage, continuous with the main ankle articular surfaces face each other. Above these is a semilunar recess of synovium, about 1 cm in height. A fringe of this synovium bulges downwards to the talar surface in plantar flexion, but retracts upwards on dorsiflexion.

The lateral malleolus contrasts with the medial malleolus in that —

i. it is about 1 cm longer. Whereas the lateral malleolus reaches down the whole height of the body of the talus the medial malleolus only reaches halfway.

ii. it is narrower.

iii. it is about 1 cm posterior to the medial malleolus, leaving an anterolateral deficiency in the mortise.

iv. it is slightly mobile at the inferior tibiofibular joint and therefore more resilient to strain.

v. only a portion of its talar-facing surface is articular.

vi. it is formed by the whole distal epiphysis of its bone rather than just a projection of it.

vii. it receives its blood supply chiefly from below, through the digital fossa. The medial malleolus receives its main blood supply from above. Non-union of the lateral malleolus is less common than of the medial.

351. What are the features of the articular surface of the talus at the ankle ?

i. The **superior** articular surface of the talus is known as the **trochlea** of the ankle. It is concave transversely and convex anteroposteriorly, reciprocating the contour of the plafond of the tibia. It is wedge-shaped to a variable degree, the posterior width ranging from 0–6 mm (average about 2.5 mm) less than the anterior. The trochlea has the form of a segment of a frustum of a cone, the apex of which is directed medially, with its axis passing obliquely close to the tips of the malleoli. Only two thirds of the trochlea is covered by the plafond in any one position. The posterior part of the lateral edge of the trochlea is bevelled slightly, abutting the deep part of the posterior inferior tibiofibular ligament. The anterolateral part of the surface can abut a low band of the anterior inferior tibiofibular ligament, which, if it becomes swollen, can impinge on and damage the articular cartilage.

ii. The **lateral** articular surface facing the lateral malleolus is relatively large and triangular, in profile forming an arc of about 100°. It is concave vertically so that its inferior apex projects a little outwards as the lateral process of the talus. It may be flat or slightly convex anteroposteriorly.

iii. The **medial** articular surface facing the medial malleolus is flat, relatively small, and comma-shaped. The tail of the comma lies posteriorly and the head anteriorly. The surface usually projects forwards a short way onto the lateral side of the neck of the bone.

352. What structures resist valgus and lateral displacement of the talus ?

Valgus and lateral displacement of the talus is resisted by —

i. The **buttress** of the lateral malleolus.

ii. The anterior, posterior and interosseous ligaments of the inferior tibiofibular joint and the interosseous membrane, holding the lateral buttress to the tibia. It is normally not possible to separate the lateral malleolus from the tibia by more than 1–2 mm, by forceful manual pressure.

iii. The tie of the medial or **deltoid ligament** of the ankle. It consists of superficial and deep components. The **superficial** part is a strong, wide, multifascicular, triangular lamina continuous with the anterior and posterior aspects of the capsule of the joint. The tip of the medial malleolus presents two prominences or colliculi, the anterior colliculus descending slightly lower than the posterior one. The colliculi are separated by a shallow notch. The upper apical attachment of the superficial part of the deltoid ligament is attached to the two colliculi and the intervening notch, and to the anteromedial margin of the plafond of the tibia. The base of the ligament fans out to attach, from front to back, to the dorsomedial aspect of the navicular, the dorsomedial aspect of the neck of the talus, the supero-medial calcaneonavicular (spring) ligament, the medial border of the sustentaculum tali, and the medial tubercle of the posterior process of the talus. The **deep** part of the ligament is its strongest component. From the colliculi and the notch it descends downwards, backwards and laterally into an oval elevation under the tail of the comma-shaped articular surface on the medial side of the talus, and to the medial tubercle of the posterior process of the talus. The deltoid ligament is so strong that when it is severely strained it is more likely to pull off the medial malleolus than rupture.

353. What structures resist varus and medial displacement of the talus ?

Varus and medial displacement of the talus is resisted by —
    i. the **buttress** of the medial malleolus.
    ii. the tie of the **lateral ligament** of the ankle joint. It consists of three distinct components. The anterior component is the **anterior talofibular ligament**, which is flat and quadrilateral, and made up of two separate bands, between which passes a branch of the peroneal artery into the talus. Blending with the capsule of the ankle joint, it extends from the lower part of the anterior border of the lateral malleolus forwards and medially to the body of the talus just in front of the lateral articular surface. It is the weakest component of the lateral ligament. It is horizontal when the foot is plantigrade and nearly vertical in extreme plantar flexion. The middle component is the **calcaneofibular ligament** which restrains the talus indirectly, bypassing it to attach to the calcaneus. It is cord-like, and discreet from the capsule. It arises from the lower part of the anterior border of the lateral malleolus, but not from the tip of the bone, which is free of ligamentous attachment. It passes downwards, backwards and medially, with a variable degree of obliquity, to attach to a tubercle on the lateral surface of the calcaneus, above and behind the peroneal tubercle. The posterior component is the **posterior talofibular ligament**. A thick band, it is the strongest of the three, and, like the anterior component, blended with the capsule. It passes nearly horizontally backwards and medially from the lower part of the digital fossa of the lateral malleolus to the lateral tubercle of the posterior process of the talus. In contrast to the medial ligament, if the lateral ligament is severely strained, it is more likely to rupture one or more of its components than pull off a large fragment of the lateral malleolus.

354. What structures resist posterior displacement of the talus?

Posterior displacement of the talus is resisted by —
    i. the slight posterior narrowing of the mortise.
    ii. the ligaments maintaining the integrity of the inferior tibiofibular joint.
    iii. reinforcement of the posterior rim of the mortise by the deep part of the posterior inferior tibiofibular ligament, which forms a transverse labrum.
    iv. reinforcement of the posterior capsule by the floors of the fibrous sheaths of the tendons of flexor hallucis longus and flexor digitorum longus.
    v. a posterior ligamentous sling. The medial side of the sling is made up of the deep part of the medial ligament, and the posterior tibiotalar

component of the superficial part of the medial ligament. The lateral side of the sling comprises the posterior talofibular component of the lateral ligament.

355. What structures resist anterior displacement of the talus?

Anterior displacement of the talus is resisted by an anterior ligamentous sling formed by the anterior talofibular component of the lateral ligament, and the anterior tibiotalar, tibionavicular and tibioligamentous components of the medial ligament, of the ankle joint. The anterior rim of the mortise has no reinforcement by ligamentous fibres, as does the posterior rim, and the tendons anterior to it do not run in fibrous tunnels. The superior extensor retinaculum is above the level of the ankle joint, and the inferior retinaculum is attached by its superior band only to the medial malleolus.

356. What movements take place at the ankle joint ?

The talus has no muscle or tendon attached to it. Its movements are therefore entirely passive, directed only by the pressure of surrounding bones. Starting from the anatomical or plantigrade position (0°), the talus can dorsiflex some 20° and plantar flex some 50°. The linear axis of these movements passes up to 1 cm below the tips of the malleoli, and is thus directed laterally, inferiorly at an angle of about 10° to the transverse plane, and posteriorly about 30° to the coronal plane, though there is some variation between individuals. In clinical practice the axis of the ankle joint can be represented by a line between the tips of the fingers pressed into the foot under the malleoli. From full plantar flexion to full dorsiflexion the talus is also said to rotate laterally, on a vertical axis, some 4°. If the relaxed foot is passively stressed forwards on the leg the talus can normally slide some 5 mm in the mortise. If the foot is stressed into varus the talus can tilt up to about 20–25°. The talus can displace the lateral malleolus laterally no more than 1–2 mm without injuring the syndesmosis.

357. What radiographic features can be used to assess the normal alignment of the ankle joint ?

The usual X-ray films of the ankle comprise an anteroposterior view, taken with the foot pointing directly forwards (i.e. in neutral rotation) with respect to the film, and a lateral view. **In adults** the following radiographic features can be used to assess the normal alignment of the ankle joint —

    a) On the anteroposterior film —

        i. the medial joint space, between the medial malleolus and the talus should be continuous with, and of the same width as, the horizontal joint space between the plafond and the trochlea.

        ii. the anterior tibial tubercle overlaps almost two thirds of the width of the fibula. The degree of tibial overlap diminishes with medial rotation of the ankle relative to the film until at 45° the overlap is hardly present.

iii. the digital fossa of the lateral malleolus is not visible, but becomes so if the ankle is medially rotated with respect to the film.

iv. a vertical, relatively translucent strip is visible between the posterior tubercle of the tibia and the medial cortex of the fibula, called the interosseous space. It widens if the syndesmosis is ruptured.

v. if passively stressed, the angle of varus tilt of the trochlea surface on the mortise should not be more than about 20–25°.

b) On the lateral film the surfaces of the plafond and the trochlea should be parallel. If the talus is displaced forwards the posterior part of the joint space opens into a wedge shape, and the distance between the anterior margin of the mortise and the upper margin of the talar articular surface for the navicular is increased in comparison with the normal side. If passively stressed, the anterior shift (anterior drawer) should not be more than about 5 mm.

The lateral joint space is commonly seen fully only on the oblique view, taken in 30–40° of medial rotation. **In children** the anterior and posterior tubercles of the tibia are not visible before the age of about 8, so that the interosseous space appears broader. After the age of 3–4 years the fibular growth plate is an even translucency, level with the horizontal joint space; before this the growth plate is a little lower.

358. What is an os trigonum ?

The posterior process of the talus presents a small medial and a larger lateral tubercle, separated by a sulcus for the tendon of flexor hallucis longus and its sheath. The lateral tubercle can vary in size from being less than noticeable to a large, prominent trigonal process. The inferior surface of the trigonal process is part of the posterior talocalcaneal articular surface. About one in ten trigonal processes show some degree of separation from the body of the talus, and in about 5% of subjects a completely separate ossicle, the os trigonum, is found. The condition is most commonly bilateral.Its inferior surface is articular. Its talar surface is joined by fibrous tissue, fibrocartilage or cartilage. The posterior talofibular ligament is attached to its posterior surface. It develops from a separate small centre of ossification which appears at about the age of 8 years. A fracture of the trigonal process (Shepherd's fracture) needs to be distinguished from an os trigonum.

359. How are the tendons inserted that dorsiflex the ankle ?

i. The tendon of **tibialis anterior** is inserted into a tubercle of the base of the first metatarsal, at the junction of its medial and inferior surfaces, and into the medial cuneiform, along the inferior and posterior margins of a smooth oval surface occupying the anteroinferior angle of its medial surface; the remainder of this latter surface is covered by a bursa between it and the tendon. Tibialis anterior is the strongest dorsiflexor of the ankle. Its insertion is close to the axis of inversion and eversion so that it can act as either a weak invertor or evertor, according to the position of the foot in relation to this axis at the moment of contraction. It is only a weak adductor of the forefoot.

ii. The tendon of **extensor digitorum longus** divides into two, deep to the superior extensor retinaculum, and then into four as the tendons pass out from deep to the lateral stem of the inferior extensor retinaculum. They pass to the second to the fifth toes. Those to the second to the fourth toes are joined laterally by the tendons of extensor digitorum brevis. The combined tendons are inserted into the phalanges in the same manner as the extensor tendons to the fingers (see question 241). Extensor digitorum longus is said to have only about a third of the power of the tibialis anterior in dorsiflexion of the ankle. It helps also to pronate the forefoot.

iii. The tendon of **extensor hallucis** is inserted into the dorsum of the base of the distal phalanx of the big toe. It is held over the proximal phalanx by the extensor hood, contributed to medially by the tendon of abductor hallucis and laterally by the tendon of adductor hallucis. It is said to have half the power of extensor digitorum longus in dorsiflexion of the ankle.

iv. The tendon of the variably developed **peroneus tertius** has a fan-like insertion into the upper surface of the proximal part of the fifth metatarsal. If it is well developed it is about equal in power to extensor hallucis longus in dorsiflexion of the ankle. It helps pronate the forefoot. It may be absent.

360. How are the tendons inserted that plantar flex the ankle ?

i. Most of the power (said to be 90%) of plantar flexion of the ankle is provided through the **tendon Achilles**, the combined tendon of insertion of the gastrocnemius and soleus. The fibres of the tendon twist through about 90° as they descend so that most of the fibres of the lateral side of the tendon at its insertion originate from the gastrocnemius, and most of those on the medial side from the soleus. It inserts into the smooth middle third of the posterior surface of the calcaneus. This area of insertion is demarcated above by a slight groove from another smooth surface, which is separated from the tendon by a bursa. It is demarcated below by an irregular ridge from a roughened area which is continuous downwards over the tuberosity of the calcaneus, to the undersurface of the heel.

ii. The tendon of **flexor hallucis longus** inserts into the plantar surface of the base of the distal phalanx of the big toe.

iii. The tendon of **flexor digitorum longus** divides in the sole into four diverging tendons. As it does so it receives slips from the tendon of flexor hallucis longus on its medial side and the insertion of flexor digitorum accessorius on its lateral side. The diverging tendons give origin to the lumbricals, and insert in the manner of the flexor tendons of the fingers into the plantar surfaces of the bases of the terminal phalanges of the second to the fifth toes (see question 239).

iv. The tendon of **tibialis posterior** has a complex insertion. It contains a bony or cartilaginous sesamoid as it passes under the spring ligament and divides into three components. An **anterior** component continues distally into the tuberosity of the navicular and the adjoining under surface of the medial cuneiform. A **middle** component continues deeply into the sole breaking up into slips that insert into the intermediate and lateral cuneiforms and the cuboid, and into three metatarsal slips which further split to insert into the bases of the four outer metatarsals. A **posterior** or recurrent component turns laterally and backwards into the anterior aspect of the sustentaculum tali. Tibialis posterior is also a powerful invertor of the hindfoot and a supinator of the forefoot.

v. The tendon of **peroneus longus**, after passing deep to the superior peroneal retinaculum behind the lateral malleolus, deep to the inferior peroneal retinaculum below the peroneal tubercle of the lateral surface of the calcaneus, and through the plantar tunnel which grooves the cuboid, inserts into the lateral tubercle of the base of the first metatarsal. Slips may pass to the adjacent part of the lateral surface of the medial cuneiform, to the base of the second metatarsal, and into the first dorsal interosseous muscle. A sesamoid forms in the tendon as it turns into its plantar tunnel, which articulates against a facet on the cuboid. It is a strong evertor of the hind foot and a pronator and abductor of the forefoot.

vi. The tendon of **peroneus brevis**, after passing deep to the superior peroneal retinaculum, and deep to the inferior peroneal retinaculum above the trochlear process, inserts into the styloid process of the fifth metatarsal. It is also a powerful evertor of the hindfoot, and pronator and abductor of the forefoot.

361. What is the blood supply of the talus ?

The **arteries from which branches supply the talus** are —

i. the dorsalis pedis artery, which is a continuation of the anterior tibial artery distal to the line of the ankle joint.

ii. the posterior tibial artery.

iii. the peroneal artery.

About two thirds of the entire talus and about three quarters of its body are covered by articular cartilage, leaving restricted areas for the entry of blood vessels. These areas comprise —

i. the superior aspect of the neck.

ii. the inferolateral aspect of the neck, which forms the roof of the sinus tarsi.

iii. the roof of the tarsal canal.

iv. beneath the facet for the medial malleolus.

v. the posterior process.

The **arterial branches which penetrate the talus** comprise —

i. those which enter the superior aspect of the neck. They are derived from arteries which arise in sequence from the anterior tibial and dorsalis pedis arteries. These are, from proximal to distal, the medial and lateral malleolar arteries, medial and lateral talar arteries, and recurrent branches of medial and lateral tarsal arteries.

ii. those which enter the inferolateral aspect of the neck. They arise from the lateral tarsal rete, which is formed by anastomosing branches of the perforating branch of the peroneal artery, and branches of the lateral malleolar, lateral talar and lateral tarsal recurrent arteries.

iii. those which enter the roof of the tarsal canal. They arise from an anastomosis between the artery of the tarsal canal and the artery of the tarsal sinus. The artery of the tarsal canal arises from the posterior tibial artery about 1 cm before it divides into medial and lateral plantar arteries and passes forwards and laterally between the tendons of flexor hallucis longus and flexor digitorum longus. The artery of the tarsal sinus can arise from any of the branches to the lateral tarsal rete.

iv. the artery which enters the medial surface of the talus, called the deltoid branch. It arises from the artery to the tarsal canal and passes into the bone between the deep and superficial parts of the medial (deltoid) ligament of the ankle.

v. those which enter the posterior process of the talus. They arise from the posterior tibial and peroneal arteries.

In order of importance the arteries of supply to the body of the talus are —

i. the anastomotic branches of the tarsal canal, which supply the middle half to two thirds, and possibly the whole, of the body.

ii. the deltoid branch, which supplies the medial third and probably more of the body.

iii. the arteries from the tarsal sinus which can supply an eighth to a half of the lateral side of the body.

iv. the superior neck branches which supply a relatively small area of the body, but can anastomose with the artery of the tarsal canal.

v. the branches to the posterior process which usually supply only a limited area, but can sometimes anastomose with the artery of the tarsal canal.

Knowledge of the blood supply of the talus is important in understanding the effects of injury. The talus is subject to fractures of the neck, and dislocation at both the ankle and subtalar joints. The blood supply to the body depends primarily on the integrity of the branches from the tarsal canal, passing upwards and backwards into it. In nearly half of cases in which the neck of the talus is fractured and the subtalar joint is dislocated avascular necrosis of the body of the bone occurs; when the ankle is dislocated as well nearly all tali show these changes.

362. What is the tarsal tunnel ?

The tarsal tunnel is the passage for the tibial nerve and posterior tibial vessels, and the deep crural tendons (from medial to lateral tibialis posterior, flexor digitorum longus and flexor hallucis longus), as they curve from behind to below the medial malleolus, and then forward to the porta pedis (entrance into the sole), above the abductor hallucis.

Superficially the tunnel is covered from above downwards by —

i. the lower part of the deep transverse fascia of the leg. This lies between the deep crural muscles and the soleus and extends from the medial border of the tibia and medial malleolus, and the fibrous sheath of the tibialis posterior tendon, to the posterior border of the fibula and lateral malleolus and the sheath of the peroneal tendons.

ii. the flexor retinaculum, continuous proximally with the deep transverse fascia, where it fuses with the deep fascia of the leg. It is triangular. The apex is attached to the medial malleolus. The base is attached to the upper border of the abductor hallucis, which it splits to enclose, and to the medial part of the tuberosity of the calcaneus.

Deeply, the floor of the tunnel consists from above downwards of —

i. the poster;or aspect of the distal tibia and medial malleolus.

ii. the posterior part of the medial surface of the talus and its posterior process with its two tubercles limiting the groove for the flexor hallucis longus.

iii. the sustentaculum tali and the concave medial surface of the calcaneus, which is covered by the origin of flexor digitorum accessorius.

The three tendons run in their fibrous sheaths attached to the floor of the tunnel. In the upper part of the tunnel the posterior tibial vessels and tibial nerve lie together between the sheaths of flexor hallucis longus and flexor digitorum longus, the artery medial to the nerve. In 90% of cases the nerve divides into medial and lateral plantar nerves within 1 cm of a line drawn from the tip of the medial malleolus to the lower part of the back of the heel. The nerve divides proximal to the artery. In the lower part of the tunnel, as the tendon of flexor hallucis longus approaches that of flexor digitorum longus, the neurovascular compartment of the tunnel becomes more posterior, and is split into two by a fibrous septum with a concave-free proximal border. The medial plantar vessels and nerve enter the anterior part, and the lateral vessels and nerve enter the posterior part of the compartment, to enter the sole above the abductor hallucis. In each case the nerve is anterior to the vessels.

The tarsal tunnel is the site of entrapment of the posterior tibial nerve in conditions such as rheumatoid arthritis and injury. In dividing the flexor retinaculum to relieve pressure it must be borne in mind that the medial and lateral plantar nerves run distally in separate canals. Also the medial calcaneal branch of the tibial nerve may arise in the tunnel, and may penetrate the retinaculum, and thus be in danger if the retinaculum is incised.

363. What are the features of the articular surfaces of the subtalar joint ?

The subtalar joint is a functional concept. It comprises those articular surfaces at which the foot can rotate under the talus about an oblique, approximately anteroposterior axis. The anatomical joints comprising the subtalar joint are the talocalcaneal and the talocalcaneonavicular. The talocalcaneonavicular joint is also part of the midtarsal joint, another functional concept.

The articular surfaces of the **talocalcaneal joint** comprise —

i. the facet under the body of the talus. It is concave, somewhat quadrilateral, with its longer axis directed anterolaterally. It underlies the apex (the lateral process) of the lateral malleolar facet, and the posterior process with its medial and lateral tubercles. If an os trigonum is present the surface continues below it.

ii. the facet on the middle third of the superior surface of the calcaneus. It is convex and similarly somewhat quadrilateral. It inclines sharply downwards and forwards, at an angle of about 65° to the upper border of the posterior third of the bone.

The surfaces of the **talocalcaneonavicular joint** comprise —

a) the facets of the head of the talus. The convex surface is somewhat ovoid, extending from the undersurface of the head to cover its distal aspect. Looked at from in front its long diameter is directed downwards and medially, at an angle of about 50° to the horizontal.

b) a mosaic of structures, sometimes described as the **acetabulum pedis**, which forms a concave surface comprising —

i. the concave ovoid surface of the navicular, its long diameter also directed downwards and medially.

ii. the anterior facet for the talus lying over the anterior beak of the calcaneus, and the middle facet for the talus overlying the sustentaculum tali. These two surfaces are often joined together.

iii. the upper, concave, cartilage-covered surfaces of the inferior (spring) and lateral (part of the bifurcate) calcaneonavicular ligaments.

364. What terms are used to describe movements and changes in the shape of the foot ?

Lateral (external) and medial (internal) rotation describe movements of the leg about a vertical axis.

Abduction and adduction describe movements of the foot about a vertical axis.

Dorsiflexion (extension) and plantar flexion (flexion) describe movements about a transverse axis.

Pronation and supination describe twists of the foot about an anteroposterior axis.

Eversion and inversion describe movements about the oblique axis of the subtalar joint, being combined movements of pronation–abduction–dorsiflexion and supination–adduction–plantar flexion respectively.

A varus heel is the position it takes up when the subtalar joint is inverted.

A valgus heel is the position it takes up when the subtalar joint is everted.

365. What movements occur at the subtalar joint ?

The axis of movement at the subtalar joint passes from the inferolateral corner of the posterior surface of the calcaneus, through the tarsal canal to the superomedial aspect of the neck of the talus. Like that of pronation and supination in the forearm, it passes through the centres of curvature of the proximal (talocalcaneal) and distal (talocalcaneonavicular) joint articular surfaces. It inclines on average about 40° upwards from the horizontal and about 20° medially. Motion of the foot about this axis causes inversion and eversion. **Inversion** causes the sole to face medially. The inclination of the axis of this movement determines that inversion is also accompanied by some plantar flexion and adduction of the forefoot. **Eversion** causes the sole to face laterally, and is likewise accompanied by some dorsiflexion and abduction of the forefoot. The range of movement about the axis is about 25–40°. Inversion is brought about, with decreasing power, by triceps surae (gastrocnemius and soleus), tibialis posterior, flexor digitorum longus and flexor hallucis longus, and tibialis anterior. Eversion is brought about, with decreasing power, by peroneus longus, peroneus brevis, extensor digitorum longus, peroneus tertius, extensor hallucis longus and tibialis anterior. Tibialis anterior inserts very close to the axis of movement and can act as a weak invertor or evertor as the foot progresses into its arc of movement.

366. On what structures does the stability of the subtalar joint depend ?

As with all joints the stability of the subtalar joint depends on the geometry of the articular surfaces, and the ligaments which join them.

Extreme movement of the calcaneus on the talus is limited by **three bony 'blocks'** —

   a) eversion of the calcaneus will ultimately cause impaction —

      i. of the lateral process of the talus into the floor of the tarsal sinus supported by the lateral wall of the calcaneus below it.

      ii. of the anterosuperior process or beak of the calcaneus into the neck of the talus.

   b) inversion of the calcaneus will ultimately cause impaction of the posterosuperior part or the sustentaculum tali against the medial tubercle of the posterior process of the talus.

The **ligaments** which support the joint comprise —

i. the **talocalcaneal interosseous ligament**, a strong flat band, the fibres of which pass upwards and medially at an angle of about 45° to the horizontal, from the floor to the roof of the tarsal canal. It is independent of the capsules of the joints in front and behind. Its lateral fibres are longer than the medial. The axis of inversion and eversion passes through it.

ii. the **cervical ligament**, a strong rectangular band extending upwards at an angle of about 45° to the long axis of the calcaneus, from a small tubercle in the medial part of the tarsal sinus, to the inferolateral aspect of the neck of the talus. It limits inversion, when it becomes nearly vertical, and eversion, when it becomes nearly horizontal.

iii. the **calcaneofibular component of the lateral ligament of the ankle**, which limits both inversion and eversion. In its most common position it lies roughly along the perimeter of the cone of movement of the subtalar joint parallel with its axis.

iv. **the medial** and **lateral talocalcaneal ligaments**. These are reinforcements of the capsule. The medial ligament extends from the posterior border of the sustentaculum tali to the medial tubercle of the posterior process of the talus and resists eversion. The lateral ligament extends from the lateral process of the talus to the nearby calcaneal surface and resists inversion

367. What movements take place at the midtarsal joint ?

Like the subtalar joint, the midtarsal joint is a functional concept. It comprises the joint surfaces by which the hind foot (talus and calcaneus) articulates with the forefoot (navicular, cuboid, and beyond), to allow more freedom of movement, particularly about an anteroposterior axis. Looked at from in front, the long diameter of the oval convex facet of the talus, for the navicular, is directed downwards and medially at an average angle of 50°. The articular surface of the navicular is significantly less than that of the talus. The facet of the calcaneus for the cuboid is saddle-shaped, convex transversely and concave vertically. Some sliding and slight twisting movement of the navicular and cuboid can thus take place at the midtarsal joint. On eversion of the hind foot the navicular slides in a downward and lateral direction on the talus and the cuboid swivels on the calcaneus, with the result that the forefoot pronates slightly on the hindfoot. The converse occurs when inversion of the hindfoot is accompanied by supination of the forefoot. The midtarsal joint also allows some of the dorsiflexion and plantar flexion which occurs during eversion and inversion respectively. Pronation and supination of the forefoot can also occur in the opposite direction to eversion and inversion. When weight is taken with the hindfoot in inversion the forefoot can pronate to keep the foot plantigrade. The medial longitudinal arch becomes exaggerated. Similarly if the hindfoot is in eversion the forefoot can supinate, when the medial arch becomes flattened. The ligaments which limit midtarsal movement are the talonavicular, calcaneocuboid and calcaneonavicular. Fractures of the beak of the calcaneus, to which some of these are attached, occur in twisting injuries of the foot.

368. What are the features of the tarsometatarsal joints ?

The combined tarsometatarsal (Lisfranc's) joints form an interlocking complex between the block formed by the cuboid and cuneiforms, and the bases of the metatarsals. In the coronal plane it forms a dorsolateral convexity. In the transverse plane it forms an oblique line passing posteriorly and laterally. The surface of the medial cuneometatarsal joint faces medially as well as distally. The intermediate cuneometatarsal joint lies nearly 1 cm proximal to the medial joint and about 0.5 cm proximal to the lateral joint. Therefore, the base of the second metatarsal is inset between the medial and lateral cuneiforms, making a more rigid joint. The relative rigidity of this joint is probably the reason why 'march' fractures are most commonly found to involve the second metatarsal. The lateral cuneiform also projects distal to the cuboid articular surface. The surface of the lateral cubometatarsal joint faces laterally as well as distally. The joints are held by dorsal, plantar and interosseous ligaments. The tarsometatarsal joints allow some **dorsiflexion and plantar flexion**. During pronation and supination the first and fifth metatarsals can rotate a little axially at their respective joints; a very little **abduction** and **adduction** can also occur. When pronation takes place at the midtarsal joint the medial cuneometatarsal joint extends and pronates and the lateral cubometatarsal joint flexes and pronates. When supination occurs the medial cuneometatarsal joint flexes and supinates and the lateral cubometatarsal joint extends and supinates.

369. What is the tarsal sinus and what does it contain ?

The tarsal sinus is the concavity on the anterior part of the upper surface of the calcaneus, which is overhung a little medially by the neck of the talus. It lies in front of the talocalcaneal joint and behind the calcaneocuboid joint. It is bounded laterally by a bony crest. Medially is the lateral boundary of the talocalcaneonavicular joint. Posteromedially from the sinus the tarsal canal passes obliquely to the medial side of the foot, between the talocalcaneal joint behind, the talocalcaneonavicular joint in front, the calcaneus below and the talus above. A group of foramina or sometimes one large foramen in the posterior part of the sinus leads into the antrum of the calcaneus (see question 370).

The tarsal sinus contains parts of—

    i. the **bifurcate ligament**, which arises by its stem from the anteromedial corner. Its medial limb is the lateral calcaneonavicular ligament and its lateral limb is the medial calcaneocuboid ligament.

    ii. the dorsolateral calcaneocuboid ligament, which arises from the anterolateral corner.

iii. the **cervical ligament**, which arises from the anterior part.

iv. the extensor digitorum brevis, which arises from the lateral part.

v. roots of the lateral stem of the inferior extensor retinaculum, which attach on both sides of the extensor digitorum brevis.

vi. a rete of anastomosing arteries, which give rise to branches which enter the superolateral aspect of the neck of the talus, and also to an artery which anastomoses with the artery of the tarsal canal, an important source of blood for the body of the talus.

The area is often swollen and tender after sprains of the foot.

370. What appearances of the calcaneus should be noted on a lateral view X-ray film ?

a) The **trabecular systems** of the calcaneus can be recognised —

i. the anterior tibial system continues the line of trabeculae downwards and backwards from the front of the tibia, across the talus to the talocalcaneal joint, and then downwards and backwards into the posterior aspect of the calcaneus.

ii. the trabeculae in line with the Achilles tendon form a dense system radiating into the posterior tubercles.

iii. anterior trabeculae radiate downwards and forwards from the tarsal sinus to the calcaneocuboid joint and the distal plantar surface.

iv. trabeculae radiate from the midplantar aspect upwards and backwards to the posterosuperior aspect and upwards and forwards to the calcaneocuboid joint.

v. a relatively porotic central area, neutral zone or triangle is outlined by the trabeculae of i, iii and iv above and is sometimes called the **antrum of the calcaneus**. If prominent it can be mistaken for a pathological condition. It lies under the pointed edge of the lateral process of the talus and is vulnerable to fracture through it.

b) **Bohler's angle** is important in recognition of depressed fractures of the calcaneus. It is formed by tracing lines from the posterosuperior edge of the posterior articular surface backwards to the posterosuperior angle and forwards to the anterosuperior angle of the outline of the bone. The angle of inclination of the anterior line to the posterior line is Bohler's angle and averages about 30–35°. It can be eliminated or even reversed in depressed fractures.

c) A spur of bone commonly projects forwards into the sole from the posteromedial tubercle. It is not of itself a cause of symptoms.

d) The anterosuperior angle of the bone may be prominent. This part may be the site of a fracture that can easily be overlooked.

e) The secondary centre of ossification over the posterior surface is histologically an epiphysis rather than an apophysis. It is a curved disc which ossifies from irregular centres. On a lateral view irregular density of the disc is sometimes asserted to be pathological, if pain is present in the heel.

371. What are the arches of the foot, and how are they supported ?

A medial longitudinal arch of the foot is described as consisting of the calcaneus, talus, navicular, cuneiforms and the medial three metatarsals. A lateral longitudinal arch is described as consisting of the calcaneus, cuboid and lateral two metatarsals. It is probably more useful, however, to consider the skeleton of the foot as consisting essentially of five sagittally aligned arcuate units. They are fused posteriorly at the calcaneus, but spread out anteriorly through the metatarsals, the heads of which all bear weight. The medial units reach higher than the lateral, and so form the obvious medial arch of the bare weight-bearing foot. The lateral units are low and do not form a clinically obvious arch. The so-called transverse arch is the concavity under the medial arch, and at most can only be considered to form half an arch, made up of the metatarsal bases, cuneiforms, navicular and cuboid.

The shape of the bones of the foot play no significant part in the maintenance of the structure of the arches. Instead the arches rely primarily on ligaments and the plantar aponeurosis for their support. The ligaments act as ties between the individual components of the arches. They comprise the long and short plantar ligaments, the interosseous talocalcaneal ligament, the spring ligament and other capsular reinforcements. The plantar aponeurosis acts passively as a tie between the ends of the longitudinal arches. It also functions actively as a windlass. As weight is taken on the forefoot and the toes are dorsiflexed the distal attachments of the aponeurosis, to the plantar plates of the metatarsophalangeal joints and the flexor sheaths of the digits, are 'wound up around the drum' of the metatarsal heads, tightening the aponeurosis, shortening the gap between the anterior and posterior pillars of the longitudinal arches, and increasing their height and rigidity.

Under static pressure, at least, **muscles seem to play no part** in the support of the arches. The leg muscles are electromyographically inactive. It is only when quite heavy loads are applied that they contribute by active contraction. Loss of muscle power, as for instance following a sciatic nerve paralysis, does not lead inevitably to collapse of the arch. Foremost among the muscles which contract in support of weight taken through the arch seems to be the flexor hallucis longus (passing under the sustentaculum tali), flexor digitorum longus and tibialis posterior. Flexor hallucis longus sends slips to the tendons of the flexor digitorum longus, chiefly to the second and third toes. Tibialis anterior, abductor hallucis and flexor digitorum brevis also seem to make some contribution.

372. What radiological features demonstrate the normal alignment of a child's foot ?

On a lateral projection, in the erect, weight-bearing position, and centred on the midfoot, an X-ray film of a child's foot shows that —

i. the axis of the talus points downwards approximately in line with the first metatarsal.

ii. the axis of the calcaneus points slightly upwards.

iii. the angle between the axes of the talus and the calcaneus (**the talocalcaneal angle**) is about 30–40°.

On a dorsoplantar projection backwards at 45° to the plane of the sole, centred on the hind foot, and with the leg tilted back —

i. the axis of the talus points at an angle of between 5–15° to the medial side or the line of the first metatarsal. This is known as **the talometatarsal angle**.

ii. the axis of the calcaneus points in approximately the line of the fourth metatarsal.

iii. the talocalcaneal angle is about 40°.

There is some variation in the two talocalcaneal angles, but together they should add up to at least 40°. In congenital talipes equinovarus both talocalcaneal angles are reduced, the axes of the talus and calcaneus pass well away from the line of the metatarsals, and the axis of the calcaneus points downwards and forwards even if dorsiflexion is stressed.

373. What are some clinically important features of the metatarsophalangeal joint of the hallux ?

The metatarsophalangeal joint of the hallux has the same basic structure as that of a metacarpophalangeal joint of a finger(see question 237), in the general shape of its articular surfaces, and the form of its capsule, plantar plate and collateral ligaments. The plantar aspect of the head of the first metatarsal differs, however, in having a median crest separating two grooved surfaces applied to the two prominent sesamoids in the plantar plate of the joint, which guide their longitudinal excursions. The tendon of flexor hallucis longus runs in a fibrous tunnel in a groove under and between the sesamoids.

In standing, with the foot plantigrade, twice as much weight is transmitted to the head of the first metatarsal, through its sesamoids, as to any other metatarsal. The metatarsophalangeal joint of the hallux is subject, therefore, to considerable strain, especially during exercise, and is the joint of the foot most liable to degenerative changes.

The angle between the first and second metatarsals (as viewed on a dorsoplantar X-ray film) is not normally greater than about 8°, and similarly the valgus angle between the hallux and the first metatarsal at the metatarsophalangeal joint is not normally more than 15–20°.

Valgus deformity of the hallux is resisted only by the abductor hallucis and the medial collateral ligament of the metatarsophalangeal joint. Such a valgus-deforming force is provided by the adductor hallucis and to some extent by the oblique course of the flexor hallucis longus and the extensor hallucis longus tendons.

For nearly half the walking cycle the joint is dorsiflexed, reaching about 60° at push-off. It is the most effective of the metatarsophalangeal joints in tightening the plantar aponeurosis in the fashion of a windlass, helping to enhance the rigidity of the foot in this part of the cycle.

Hallux valgus is associated with an increase in the angle between the first and second metatarsals, in which the medial part of the head of the first metatarsal becomes prominent. A bursa may develop over it and form a bunion. The hallux drifts into valgus and pronates. The abductor hallucis loses its effectiveness and the valgus forces acting on the big toe predominate, increasing the deformity. The plantar plate with its sesamoids drifts laterally with the proximal phalanx, to which it is firmly attached, flattening the median crest. Osteoarthritis of the joint develops.

THE SKULL

374. What are the parts of the skull ?

The skull is the skeleton of the head.
The mandible is the lower jaw.
The skull, less the mandible, is the cranium.
The cranium, less the skeleton of the face, is the calvaria.
The calva or skull cap is the upper part of the calvaria, sawn off to examine the interior.
The vault is the concave surface of the roof of the calvaria, coextensive with the calva.
The base is the floor of the calvaria.

375. What does the skull do ?

Most obviously, the skull protects the brain from external impact, and the stresses caused by the attached masticatory and axial musculature. By enclosing a fluid bed of cerebrospinal fluid, and by giving attachment to the falx cerebri and falx cerebelli, it protects the brain from internal impact. It isolates the cerebral circulation from external pressures. It maintains an airway through the nasal cavities. It protects the eyes and provides the physical support necessary for binocular vision. It provides a fixed relationship between the parts of the vestibular apparatus, and protects the cochleae. It provides the skeleton of, and the muscular attachments for a strong masticatory apparatus. It provides attachment for the pharynx and helps suspend the larynx. It provides attachment for the axial musculature of the neck.

376. How do the bones of the adult cranium join together ?

i. The majority of bones of the cranium articulate by **sutures**, one of the three forms of fibrous joint, the other two being the gomphoses of tooth sockets, and syndesmoses. While some sutures are relatively straight, the majority are not. They may be serrated, to resist shearing stresses, or have peg- and socket-like interlocking projections, which resist distraction as well.

The coronal, sagittal and lambdoid sutures show this structure. Some sutures are bevelled, so that one surface overrides another. Such surfaces also have interdigitating irregularities, more prominent in the outer table than the inner table. The squamous temporal bone has this form of suture with the parietal. The facing edge of each bone is covered by periosteum, which shows the usual distinction between deep and superficial fibrous layers. Between facing fibrous layers of periosteum is loose fibrous tissue. Small veins in this tissue communicate with scalp veins, diploic veins and intracranial venous sinuses. The outer and inner fibrous layers continue across the joint to form outer and inner sutural ligaments.

ii. Some bones of the cranium articulate by cartilaginous remains of the chondrocranium, a form of **synchondrosis**. The basal parts of the sphenoid and occipital bones are joined by cartilage until the early twenties. The foramen lacerum, the site of junction between the sphenoid and temporal bones, remains filled with fibrocartilage. The sutures between the petrous temporal and the greater wing of the sphenoid, and between the petrous temporal and the occipital bone are wider than most other sutures, and contain some fibrocartilage. They are sometimes described as **fissures** rather than sutures.

iii. Beyond thirty years of age some sutures begin to ossify, on the inner aspect of the calvaria before the outer aspect, so that the adjoining bones articulate by **synostoses**. In the calvaria fusion involves, for example, the sagittal suture, the middle of the coronal and lambdoid sutures, and the sphenofrontal and occipitomastoid sutures. In the face the sutures of the palate are the first to fuse. The zygomaticofrontal and nasofrontal sutures tend to remain open.

THE CALVA

377. What bones make up the calva ?

The calva is made up of a large part of the frontal, most of the parietal, and the upper part of the squamous occipital bones.

378. What sutures can be seen in the calva ?

The calva presents the following sutures —

the **coronal** suture, between the frontal and parietal bones.

the **sagittal**, between the parietal bones.

the **lambdoid**, between the parietal and occipital bones. This suture is the most common site of small ossicles known as sutural or Wormian bones. Goethe's ossicle, or the Inca bone, is a larger one sometimes found at the junction of the sagittal and lambdoid sutures. It is one of a variety of possible separate small bones which arise from the multiple centres of membranous ossification of the squamous part of the occipital bone.

Occasionally a **metopic** suture between the two halves of the frontal bone persists.

The point of junction of the sagittal and coronal sutures is called the **bregma**. The point of junction of the sagittal and lambdoid suture is called the **lambda**.

379. Identify the temporal lines.

The temporal lines arise on each side anteriorly as a single ridge from the posterior border of the zygomatic process of the frontal bone. As each curves upwards and backwards onto the calva it divides into two less distinct markings, enclosing a smooth area between them on the parietal bone. The upper line fades away. The inferior line continues backwards on the parietal, and then downwards, off the calva, onto the squamous temporal bone, where it then curves forwards to become continuous with the supramastoid crest. The upper line and smooth area are for attachment of the temporal fascia. The lower line marks the upper limit of the origin of the temporalis muscle.

380. Identify a parietal foramen.

A parietal foramen may be present on each side, close to the sagittal suture, about 3.5 cm anterior to the junction of the sagittal suture with the lambdoid suture. A small emissary vein passes through it from the superior sagittal sinus to the occipital veins.

381. Identify the frontal crest.

Within the skull the frontal crest projects backwards from the anterior midline for attachment of the falx cerebri.

382. Identify the sagittal sulcus.

The sagittal sulcus begins anteriorly where the upper end of the frontal crest divides into two low ridges, which demarcate the groove for the superior sagittal sinus. The sulcus extends backwards under the sagittal suture to the midline of the occipital bone.

383. Identify the granular foveolae.

The granular foveolae are the several small depressions, more apparent in elderly skulls, near the superior sagittal sulcus. They are the site of clusters of arachnoid granulations protruding into the sinus or into one of its adjacent lacunae.

384. Identify the meningeal vascular markings.

The meningeal vascular markings are the grooves which ascend the internal surface of the calva, and correspond to the anterior (frontal) and posterior (parietal) branches of the middle meningeal vessels. The anterior branches pass up just behind the coronal sutures. The markings are made by the veins. They are in fact venous sinuses, formed between the layers of dura. They communicate inferiorly with the cavernous sinus, or with the pterygoid plexus through the foramen ovale or spinosum. They drain diploic channels, superficial middle cerebral veins, and veins from the meninges. The arterial branches that accompany them lie on their deeper aspect, and supply the bone and diploë more than the dura.

385. What does a section of the calva look like ?

The bone of the calva consists of inner and outer tables of cortical bone, between which is cancellous bone called diploë. The inner table is thinner than the outer. The inner table is said to be more brittle than the outer table, and to fracture more easily, sometimes even without obvious accompanying fracture of the outer table. This may be because, when a blow deforms the outer table, the inner table is subject mostly to tension rather than compression, and therefore fractures first. Exit wounds can cause more fracturing of the outer table than the inner table, when the reverse force is applied. The diploë contains red marrow.

386. Identify a diploic channel.

A diploic channel appears as a local deficiency in the cancellous bone between the two tables of the calva. In life it contains a diploic vein, formed of a lining endothelium, supported only by delicate elastic tissue. Diploic veins are irregular in calibre and are devoid of valves. Frontal, anterior temporal, posterior temporal and occipital diploic veins are the larger ones receiving names. They drain within the skull to dural sinuses and meningeal veins. They drain externally to pericranial veins, for example, through the foramen in the supraorbital notch to the supraorbital vein, and through the mastoid foramen to a posterior auricular or occipital vein.

387. What are the surface markings of the lobes of the cerebral hemispheres beneath the calva ?

A common point of reference on the scalp is in the sagittal plane midway between the **glabella,** which is the midpoint of the prominence between the superciliary arches, and the **inion,** which is the midpoint of the external occipital protuberance. The **bregma,** the meeting point of the coronal and sagittal sutures, and the site of the closed anterior fontanelle, is 2–3 cm anterior to this reference point. The upper end of the central sulcus lies about 1 cm behind the reference point, that is about 3–4 cm behind the bregma. It passes obliquely downwards, forwards and laterally at an angle of about 70–80° to the sagittal suture, to a point about 5 cm above the posterior root of the zygoma, just in front of the auricle. It reaches to the squamoparietal suture. Anterior to the central sulcus is the 1 cm wide precentral gyrus of the principal motor cortex of the frontal lobe. Posterior to the central sulcus is the 1 cm wide gyrus of the principal sensory cortex of the parietal lobe. The motor speech area lies roughly under the lower, lateral end of the coronal suture on the left side in right handed subjects, and on the right side in left handed subjects. The posterior part of the lateral sulcus passes backwards across the parietal bone, from the Sylvian point in the centre of the area of the pterion, just below the calva to end about 1 cm below the parietal eminence. The parieto-occipital sulcus is located opposite or just above the lambda, the point of junction of the sagittal and lambdoid sutures, about 7 cm above the inion. The parietal lobe is outlined by the midline, the central sulcus, the lateral sulcus and the parieto-occipital sulcus. Behind the parieto-occipital sulcus is the occipital lobe.

THE ANTERIOR CRANIAL FOSSA

388. What are the boundaries of the anterior cranial fossa ?

Anteriorly and laterally the anterior cranial fossa is bounded by the frontal bone. Posteriorly the fossa is demarcated, from the midline laterally, by the anterior lip or limbus of the chiasmatic sulcus, the upper margin of the posterior opening of the optic canal, the anterior clinoid process and the posterior border of the lesser wing, all of the sphenoid bone. The most lateral part of the posterior boundary is usually formed by the junction of the greater wing of the sphenoid with the orbital plate of the frontal bone, the lesser wing of the sphenoid falling a little short of the extremity of the boundary at the sphenoidal angle of the parietal bone.

389. What bones form the anterior cranial fossa ?

The bones which contribute to the anterior cranial fossa are —

the **frontal**, which forms most of the floor, and the anterior and lateral boundaries.

the **ethmoid**. The crista galli and the cribriform plate lie centrally. On each side the labyrinth, containing the ethmoidal sinuses, is roofed over by the orbital plate of the frontal bone.

the **sphenoid**. The jugum sphenoidale of the body, and the lesser wings form the posterior part of the fossa.

the **parietal**. The anterior inferior or sphenoidal angle may make a minor contribution to a posterolateral angle of the fossa.

390. What are the contents of the anterior cranial fossa ?

The contents of the anterior cranial fossa are —

The undersurface of the frontal lobes.

The olfactory bulbs and tracts.

The olfactory nerve fasciculi which lead from the olfactory nasal epithelium to the olfactory bulbs through the foramina of the cribriform plate.

The anterior cerebral arteries which pass forwards below the olfactory tracts and are joined by the anterior communicating artery. Small anterior cerebral veins accompany the arteries.

The anterior ethmoidal vessels and nerves, which run a short distance on the cribriform plate, on either side of the crista galli, give meningeal branches in the fossa, and then descend into the nose.

Branches of the posterior ethmoidal vessels which emerge through small foramina near the posterolateral angles of the cribriform plate.

391. What are the immediate relations of the floor of the anterior cranial fossa ?

Above the floor on each side are —

Orbital gyri and the gyrus rectus of the frontal lobe.

The olfactory bulb and tract below the olfactory sulcus.

The proximal course of the anterior cerebral artery and anterior cerebral vein.

Anterior and posterior ethmoidal vessels and anterior ethmoidal nerve.

The anterior communicating artery crosses the midline above the jugum sphenoidale.

Below on each side are —

The **orbit**. In contact with the undersurface of the roof are the lacrimal, frontal, supraorbital, supratrochlear and trochlear nerves. The lacrimal nerve lies above the upper border of lateral rectus. It supplies the lacrimal gland, situated below the anterolateral part of the roof in a shallow fossa, before passing on to skin. The frontal nerve and its two branches, and the trochlear nerve lie above levator palpebrae, which in turn lies above superior rectus. The upper border of superior oblique lies below the roof medially.

The **optic canal**, beneath the superior root of the lesser wing of the sphenoid, which contains the optic nerve and its dural sheath, and the ophthalmic artery.

The **frontal sinus**. It may extend posteriorly in the medial part of the orbital plate nearly as far as the lesser wing of the sphenoid near to the optic canal. It can extend upwards to the middle of the forehead and outwards towards the middle of the supraorbital margin and roof of the orbit.

Ethmoidal sinuses. Anterior, middle and posterior sinuses lie under the medial part of the orbital plate of the frontal bone, and may encroach a little into the plate laterally.

Below in the midline are —

The **two sphenoidal sinuses**, under the jugum sphenoidale. On each side a sinus may extend laterally into the roots of the lesser wing of the sphenoid partially to surround the optic canal.

The upper part of the nasal cavities below the cribriform plate.

392. What clinical effects could follow a fracture of the anterior cranial fossa ?

Following a fracture of the anterior cranial fossa there could be —

Intracranial bleeding, forming an extradural haematoma below the frontal lobe.

Intraorbital bleeding, causing subconjunctival haemorrhage without a visible posterior limit, and periorbital bruising.

Bleeding from the nose — **epistaxis**.

Loss of CSF from the nose — **rhinorrhoea**.

Blindness from injury to the optic nerve or ophthalmic artery.

Entry of air into the subarachnoid space and ventricular system — **aerocele**

Intracranial infection from the nasal cavities or air sinuses.

Loss of sense of smell — **anosmia** — from damage to the olfactory nerve fasciculi, bulbs or tracts.

THE MIDDLE CRANIAL FOSSA

393. What are the boundaries of the middle cranial fossa ?

Anteriorly the middle cranial fossa is bounded by the posterior margin of the anterior cranial fossa (see question 388). Posteriorly, the middle cranial fossa is bounded by the anterior margin of the posterior cranial fossa. This is formed centrally by the upper border, posterior clinoid processes and lateral borders of the dorsum sellae, and laterally by the superior borders of the petrous temporal bones, also called the petrous ridges. Laterally, on each side, the boundary is made up of the upper parts of the greater wing of the sphenoid and the squamous part of the temporal bone, and the inferior border of the parietal bone.

394. What bones form the middle cranial fossa ?

The bones that comprise the middle cranial fossa are —

The **sphenoid**. The chiasmatic sulcus joins the optic canals. The tuberculum sellae, pituitary fossa and dorsum sellae comprise the sella turcica in the middle of the fossa. The greater wing forms the anterior third of the fossa on each side.

The squamous part, and the anterior surface of the petrous part, of the **temporal bone** together form the posterior two thirds of the fossa on each side.

The lower edge of the **parietal** bone adjoins the squamous part of the temporal bone on each side laterally.

395. What are the contents of the middle cranial fossa ?

The middle cranial fossa contains —

The temporal lobes.

The optic nerves, chiasma, and the beginning of the optic tracts.

The oculomotor, trochlear, ophthalmic, maxillary, mandibular, abducent, greater and lesser petrosal nerves. There are sympathetic plexuses around the internal carotid and middle meningeal arteries. An external petrosal nerve may connect the sympathetic plexus around the middle meningeal artery to the ganglion of the facial nerve through the hiatus for the greater petrosal nerve.

The cavernous, intercavernous, sphenoparietal, superior petrosal, and petrosquamous venous sinuses. The middle meningeal veins are also sinuses structurally.

The internal carotid arteries in the cavernous sinuses, their subsequent division into anterior and middle cerebral arteries, and the anterior and posterior communicating arteries, forming the anterior part of the circle of Willis. Branches of the internal carotid arteries within the anterior cranial fossa also include inferior and superior hypophyseal, anterior choroidal and ophthalmic arteries. The middle meningeal arteries divide into their anterior and posterior branches within the middle cranial fossa.

The pituitary gland and its stalk.

396. Identify the optic foramen. What passes through it ?

The optic foramen is the opening of the optic canal, a tunnel about 5–10 mm long, passing anteriorly and laterally between the roots of the lesser wing of the sphenoid, to open at the apex of the orbit. It transmits the optic nerve enclosed in its sheaths of pia, arachnoid, and dura mater. The dura is adherent to the bone of the canal. At the orbital opening of the canal the dura splits to form the dural sheath of the optic nerve and the periosteum (periorbita) of the orbit. The subarachnoid space around the optic nerve is continuous with the chiasmatic cistern of the cranial cavity, so that increased intracranial pressure can affect the nerve. The ophthalmic artery and its sympathetic plexus pass through the canal, lying inferolateral to the nerve. The artery is embedded in the dura of the canal, but emerges from it in the orbit to pass medially above the nerve.

397. What are important relations of the optic canal ?

Medially lie the sphenoidal and posterior ethmoidal sinuses. The canal indents the upper lateral wall of the sphenoidal sinus. Rarely the sphenoidal sinus can extend into the roots of the sphenoid above and below the canal and thus surround the nerve.

Infection or malignancy in these sinuses could affect the nerve.

398. Identify the chiasmatic sulcus.

The chiasmatic sulcus extends transversely between the two optic canals over the dorsal aspect of the body of the sphenoid, posterior to the smooth area of the jugum sphenoidale between the lesser wings, and superior to the tuberculum sellae. Its anterior margin is called the limbus sphenoidale. The optic chiasma lies posterosuperior to it, and is rarely in direct contact.

399. Identify the foramen rotundum. What passes through it ?

The foramen rotundum passes through the anteromedial aspect or root of the greater wing of the sphenoid. It transmits the maxillary nerve to the pterygopalatine fossa.

400. Identify the foramen lacerum. What does it contain ?

The foramen lacerum is a short canal between bones, rather than through a bone. It is bounded by the body of the sphenoid, its greater wing and lingula process, and the apex of the petrous part of the temporal bone. Only a meningeal branch of the ascending pharyngeal artery, and several small emissary veins between the cavernous sinus and the pterygoid plexus and pharyngeal veins pass right through the foramen. The internal carotid artery, with its plexus of sympathetic fibres and veins around it, enters the foramen posteriorly from the carotid canal of the petrous temporal. It turns upwards to enter the cavernous sinus. The greater petrosal nerve passes forwards across the trigeminal impression, below the trigeminal ganglion, and into the foramen lateral to the internal carotid artery. It joins with the deep petrosal nerve formed from the sympathetic plexus around the internal carotid artery. The two nerves join to form the nerve of the pterygoid canal, which leaves the foramen lacerum through its anterior wall, under the lingula, in the pterygoid canal, to open into the pterygopalatine fossa. Fibrocartilage, a remnant of the chondrocranium, fills the foramen between its contents.

401. Identify the foramen ovale. What passes through it ?

The foramen ovale passes through the posterolateral angle of the greater wing of the sphenoid, to the infratemporal fossa. It transmits the mandibular nerve, with the smaller motor branch of the trigeminal nerve inferior and medial to it. Also passing through the foramen are the accessory meningeal artery, usually the lesser petrosal nerve, and a network of small emissary veins connecting the cavernous sinus with the pterygoid plexus of veins.

402. Identify the foramen spinosum. What passes through it ?

The foramen spinosum is the small round foramen lying posterolateral to the foramen ovale in the posterior angle of the greater wing of the sphenoid. It transmits the middle meningeal artery, a meningeal vein and the nervus spinosus (the meningeal branch of the mandibular nerve).

403. Identify the two small foramina on the anterior surface of the petrous part of the temporal bone. What passes through them ?

The larger, more medial foramen on the anterior surface of the petrous part of the temporal bone is **the opening of the canal for the greater petrosal nerve**, which emerges from the geniculate ganglion of the facial nerve in its canal within the petrous temporal bone. The greater petrosal nerve passes along the groove leading from the foramen towards the trigeminal impression, and then enters the foramen lacerum.

The smaller, more lateral foramen, is **the opening of the canal for the lesser petrosal nerve**, which emerges from the tympanic plexus on the promontory of the tympanic cavity, and passes through the foramen ovale, to the otic ganglion. It may pass through a foramen innominatum.

404. Identify the impression for the trigeminal ganglion.

The impression for the trigeminal ganglion is the gentle concavity overlaying the apex of the petrous temporal, posterolateral to the foramen lacerum and posteromedial to the foramen ovale. The trigeminal roots and the posterior part of the ganglion are partially enclosed in an evagination of dura and arachnoid from the posterior cranial fossa, which extends outwards, from under the tentorium, usually between the petrous ridge and the superior petrosal sinus, to overly the bony impression. Often the superior petrosal sinus passes below the ganglion, or splits to pass above and below it. The diverticulum of dura and arachnoid around the trigeminal ganglion is known as Meckel's cave. It is a little more extensive below the ganglion than above it, and towards the mandibular branch. The greater petrosal nerve passes below it.

405. Identify the markings of the middle meningeal vessels.

The middle meningeal artery and its companion veins pass through the foramen spinosum and then forwards and laterally onto the squamous part of the temporal, where they divide into anterior or frontal and posterior or parietal branches. The anterior branches pass onto the greater wing again and ascend to the anterior inferior angle of the parietal at the pterion, and so on towards the vertex. The posterior branches pass backwards and upwards to the lower border of the parietal anterior to its posterior inferior angle.

406. What is the pterion ?

The pterion is the meeting place, at the side of the skull, of four bones, the frontal, the parietal, the squamous part of the temporal, and the greater wing of the sphenoid. It covers an area of about 2 cm in diameter, located about 5 cm above the middle of the zygomatic arch. Another approximate surface marking for it is three fingers' breadths above the zygomatic arch and a thumb's breadth behind the frontal process of the zygomatic bone. Its importance lies in the fact that it overlies the course of the anterior branches of the middle meningeal vessels, which often, at this point, lie in a deep groove or tunnel in the inner table of the skull, and are vulnerable to rupture in head injury. The centre of the area is known as the Sylvian point, marking the site of the underlying anterior end of the lateral sulcus of the cerebral hemisphere, and the motor speech cortex on the dominant side.

407. What are the boundaries of the pituitary fossa ?

The pituitary fossa is bounded in front and below by the body of the sphenoid, with its two variably developed air sinuses. The anterior wall is limited above by the tuberculum sellae, behind the chiasmatic sulcus. Posteriorly the fossa is limited by the dorsum sellae. On either side is the medial wall of the cavernous sinus, and above, the fossa is roofed in by the diaphragma sellae, perforated by the aperture for the pituitary stalk.

408. Where is the apparatus of the ear in relation to the middle cranial fossa ?

i. The cochlea is enclosed in the part of the petrous temporal between the fundus of the internal auditory meatus medially and the medial wall of the tympanic cavity laterally. It lies under a slight concavity on the anterior surface of the petrous part of the temporal bone between the posterior ridge of the trigeminal impression anteriorly and the arcuate eminence posteriorly.

ii. The floor of the middle fossa between the trigeminal impression and arcuate eminence medially, and the squamous part of the temporal bone laterally, is formed by a thin lamina of compact bone called the **tegmen tympani**. It forms the roof, from in front to behind, of the canal for the tensor tympani above the pharyngotympanic canal, the tympanic cavity, and the aditus to the mastoid antrum.

iii. The arcuate eminence overlies the anterior semicircular canal.

409. Where does the internal carotid artery imprint the middle cranial fossa ?

The internal carotid artery enters the cavernous sinus through the foramen lacerum, and turns forward on the carotid sulcus. Anteriorly it bends upwards to the medial side of the anterior clinoid process, to pierce the roof of the sinus. The anterior end of the medial lip of the carotid sulcus may be prominent enough to form a middle clinoid process, and may be joined by a bar of bone to the anterior clinoid process, to form the posterior boundary of a caroticoclinoid foramen through which the artery passes.

410. What intracranial venous sinuses are in contact with the middle cranial fossa ?

On each side of the cranial fossa there are —

i. a **cavernous sinus**, which forms the lateral wall of the pituitary fossa. It extends between the anterior and posterior clinoid processes, and overlies the superior opening of the carotid canal. It receives ophthalmic veins anteriorly, inferior and superficial middle cerebral veins superiorly, intercavernous veins medially and meningeal veins and the sphenoparietal sinus laterally. Superior and inferior petrosal sinuses leave it posteriorly. Emissary veins leave it inferiorly through the foramen lacerum.

ii. a **sphenoparietal sinus** along the posterior margin of the lesser wing. It may receive the anterior temporal diploic vein, and drains into the cavernous sinus.

iii. a **superior petrosal sinus**, along the superior border of the petrous temporal bone. It drains from the cavernous sinus to the sigmoid sinus, passing usually above, but sometimes below, and sometimes on both surfaces of, the trigeminal cave.

iv. a **petrosquamous sinus**, on the fissure between the petrous and squamous parts of the temporal bone. It drains posteriorly into the straight sinus and anteriorly through a squamosal or postglenoid foramen to the retromandibular vein.

v. the **middle meningeal veins**.

411. What emissary veins communicate with the middle cranial fossa ?

Emissary veins in the middle cranial fossa are —

i. the ophthalmic veins, superior and inferior, which communicate posteriorly with the cavernous sinus through the superior orbital fissure. The inferior vein communicates in addition with the pterygoid plexus through the inferior orbital fissure, thereby forming a secondary communication between the ophthalmic veins and the cavernous sinus.

ii. veins passing through the foramen lacerum, and the foramen ovale or emissary sphenoidal foramen, which link the cavernous sinus with the pterygoid and pharyngeal plexuses.

iii. a plexus of small veins which accompanies the internal carotid artery from the cavernous sinus to the internal jugular vein.

iv. the inferior petrosal sinus which passes between the cavernous sinus and the internal jugular vein.

v. the petrosquamous sinus which communicates through the squamosal or posterior glenoid foramen with the retromandibular vein anteriorly, and drains into the straight sinus posteriorly. It also receives veins from the tympanic cavity.

THE POSTERIOR CRANIAL FOSSA

412. What are the boundaries of the posterior cranial fossa ?

Anteriorly, the posterior cranial fossa is bounded by the posterior border of the middle fossa (see question 393). Posteriorly the petrous ridge forms the upper lip of the upper part of the sigmoid sinus, and then continues medially and posteriorly as the upper lip of the sulcus for the transverse sinus, which forms the posterior boundary of the fossa. On the right side the sulcus for the transverse sinus is usually continuous in the midline with the sulcus for the superior sagittal sinus. On the left side the upper lip of the sulcus for the transverse sinus reaches to the internal occipital protuberance, where the sulcus usually ends. The boundary on each side can be represented on the surface of the head as level with a line passing backwards from the posterior root of the zygoma to just below the external occipital protuberance.

413. What bones contribute to the posterior cranial fossa ?

The bones that make up the posterior cranial fossa are —
   The sphenoid bone.
   The occipital bone.
   The **clivus** which is the surface of bone that slopes downwards and backwards from the dorsum sellae to the foramen magnum, formed by fusion of the posterior part of the body of the sphenoid with the basilar part of the occipital bone.
   The petromastoid parts of the temporal bones.
   The posteroinferior or mastoid angles of the parietal bones.

414. What forms a roof for the posterior cranial fossa ?

The tentorium cerebelli forms a sloping, crescentic roof for the fossa. Its peripheral border is attached on each side to the posterior clinoid process and lateral border of the dorsum sellae, and the anterolateral and posterior boundaries of the fossa. In the attached border are the superior petrosal and transverse sinuses. Its free border defines the tentorial incisure and is attached anteriorly on each side to an anterior clinoid process. In the midline, at the junction of the falx cerebelli with the falx cerebri, is the straight sinus. The tentorium supports the occipital lobes of the cerebral hemispheres.

415. What are the contents of the posterior cranial fossa ?

The contents of the posterior cranial fossa are —

Parts of the brain. The medulla, pons, cerebellum and the fourth ventricle fill the fossa. The midbrain and the anterior part of the vermis of the cerebellum occupy the incisure of the tentorium.

Cerebrospinal fluid, contained in the cisterna magna and cisterna pontis, and the cerebellopontine angle cisterns.

Several arteries. The vertebral arteries join to form the basilar artery. From these arteries arise meningeal, anterior and posterior spinal, medullary, cerebellar (anterior and posterior inferior, and superior), labyrinthine and posterior cerebral branches. A choroid plexus lies in the roof of the fourth ventricle.

Various veins. The superior petrosal and transverse sinuses, and the straight sinus are contained within the layers of the tentorium. The inferior petrosal sinuses are connected by a basilar plexus of veins on the clivus. The occipital sinus often divides into two to form a marginal sinus around the posterior part of the foramen magnum. The great cerebral vein (of Galen) drains into the straight sinus at its junction with the inferior sagittal sinus.

A number of nerves. The trigeminal, abducent, facial, vestibulo-cochlear, glossopharyngeal, vagus, cranial and spinal accessory, and hypoglossal nerves pass out of the fossa. Recurrent branches of the vagus and hypoglossal nerves supply the dura of the posterior cranial fossa, carrying fibres from the first two cervical nerves and the superior cervical sympathetic ganglion.

416. What passes through the foramen magnum ?

The spinal dura is continuous with the cerebral dura at the foramen magnum. **Within the dura** at the foramen magnum is the lower medulla. It lies in a cushion of cerebrospinal fluid enclosed in arachnoid mater. On each side the upper denticulum of the denticulate ligament attaches just within the foramen. On each side the spinal accessory nerve passes through the foramen behind, and the vertebral artery in front, of the denticulate ligament. The single anterior spinal artery passes down through the foramen in the midline anteriorly. It is formed at mid-medullary level by the union of anterior spinal branches which arise one from near the end of each vertebral artery. On each side a posterior spinal artery arises from the vertebral artery lower down, or from the posterior inferior cerebellar artery, and passes through the foramen posterolaterally, to descend as two branches, anterior and posterior to the dorsal roots of the spinal nerves. **Outside the dura** the tectorial membrane, and the apical ligament, from the axis, reach to just above the anterior margin of the foramen. Meningeal branches derived from the sinuvertebral nerves of the upper three cervical spinal nerves ascend through the anterior part of the foramen, between the spinal dura and the tectorial membrane. Communicating veins pass between the basilar plexus of veins, the marginal sinus and occipital sinus, and the veins of the internal vertebral venous plexus in the plane between the dura and periosteum.

417. Identify the jugular foramen. What passes through it ?

The jugular foramen is the large, irregular foramen, opening downwards, forwards and laterally, between the posterior border of the petrous temporal bone and the jugular notch of the occipital bone. It is usually partially divided into three parts, by bony projections on either side of a notch on the posterior border of the petrous part of the temporal bone, which forms the anterior, superior or lateral boundary of the foramen. The cochlear canaliculus opens into the apex of this notch. The inferior petrosal sinus passes through the anterior part of the foramen, as does also a meningeal branch of the ascending pharyngeal artery. The sigmoid sinus empties into the bulb of the internal jugular vein in the expanded posterior part of the foramen. A meningeal branch of the occipital artery also passes through this part of the foramen. The glossopharyngeal nerve passes through the lateral part of the intermediate portion of the foramen. It is within its own sleeve of dura. The two ganglia of the nerve are within the foramen, the upper one lying in the notch in the petrous part of the temporal bone. The tympanic branch of the inferior ganglion passes through a small foramen in the ridge of bone which separates the jugular foramen from the carotid foramen in front of it. The vagus and accessory nerves pass through the medial part of the intermediate part of the foramen. They are enclosed together within a sleeve of dura, separated from the sleeve around the glossopharyngeal nerve by the continuation of the inferior petrosal sinus to the jugular bulb. The superior ganglion of the vagus nerve lies in the jugular foramen and may often be just visible from within the cranial cavity. A meningeal branch of the superior ganglion of the vagus carries fibres derived from the first cervical and second spinal nerves in a recurrent course through the foramen. The auricular branch of the superior ganglion of the vagus, joined by a branch of the inferior ganglion of the glossopharyngeal nerve, passes behind the internal jugular vein to escape the foramen through the lateral wall of the jugular fossa.

418. Identify the internal auditory meatus. What passes through it ?

The internal auditory meatus is the large foramen in the middle of the posterior surface of the petrous temporal bone. Passing into the meatus are the facial nerve anteriorly and the vestibulocochlear nerve posteriorly. Between them is the nervus intermedius, at first more associated with the vestibulocochlear nerve, but laterally joining the facial nerve, sometimes as more than one branch. The labyrinthine artery courses below the nerves. It frequently arises from the summit of a short loop of the anterior inferior cerebellar artery which penetrates a short distance into the meatus before emerging again. A labyrinthine vein emerges from the meatus to drain into either the superior petrosal sinus or the transverse sinus. The lateral end of the canal, or fundus, is a vertical wall divided by a transverse crest into upper and lower parts. The facial nerve enters its canal through a foramen above the crest anteriorly. Branches of the vestibular component of the vestibulocochlear nerve to the utricle, and anterior and lateral semicircular canals, penetrate foramina of the superior vestibular area, behind the opening of the facial canal, and above the crest. The nerves of the cochlear component of the vestibulocochlear nerve pass through the foramina of the tractus spiralis foraminosus below the anterior part of the transverse crest, and branches of the vestibular component to the saccule pass through foramina of the inferior vestibular area below the posterior part of the transverse crest. The nerve to the posterior semicircular canal passes through the foramen singulare below and behind the inferior vestibular area.

419. Identify the hypoglossal canal. What passes through it ?

The hypoglossal canal leads forwards, upwards and laterally from a foramen situated above the anterolateral part of the foramen magnum, below the jugular tubercle. It is often partially or completely divided into two by a bony septum. Through it pass the joining rootlets of the hypoglossal nerve, and its recurrent meningeal branch which carries C1 and 2 fibres, and sympathetic fibres from the superior cervical ganglion. A branch of the ascending pharyngeal artery passes through the foramen to supply the nerve, the meninges and adjacent bone. A plexus of small veins, or a single vein also pass through the foramen, between the sigmoid sinus and the internal jugular vein.

420. What emissary veins communicate with the posterior cranial fossa ?

Emissary veins of the posterior cranial fossa pass through —

the **foramen magnum**, joining the occipital sinus, marginal sinus, and basilar venous plexus, with the internal vertebral plexus.

the **mastoid foramen**, joining the sigmoid sinus with the posterior auricular or occipital veins.

the **hypoglossal canal**, as a plexus of small veins or a single vein, joining the sigmoid sinus with the internal jugular vein.

the **condylar canal**, joining the sigmoid sinus with the suboccipital veins.

Sometimes, a foramen in the occipital protuberance, joining the confluence of sinuses with the occipital veins.

421. How does the ear relate to the posterior cranial fossa ?

Usually only a thin lamina of bone separates the upper part of the sigmoid sinus from the mastoid antrum and air cells, posing a potential hazard in the presence of infection. The aqueduct of the vestibule opens via a small slit on the posterior surface of the petrous part of the temporal bone behind the internal auditory meatus. It conducts the endolymphatic duct, which ends as the endolymphatic sac, which protrudes through the opening to lie between the two layers of dura. A small artery and vein accompanies the ductus. The cochlear canaliculus opens into the apex of the notch on the posterior border of the petrous part of the temporal bone. It provides an escape route for perilymph to the cerebrospinal fluid in the adjacent subarachnoid space. In the infant a small hollow is present above and between the internal auditory meatus and the hiatus for the endolymphatic sac, called the subarcuate fossa. It is formed by the flocculus of the cerebellum protruding a little through the ring of the posterior semicircular canal. It is barely visible in the adult.

422. What is the course of the abducent nerve in the posterior cranial fossa ?

On each side an abducent nerve diverges from the sulcus between the posterior border of the pons and the pyramid of the medulla. It passes forwards and upwards in the pontine cistern of cerebrospinal fluid, usually dorsal to the anterior inferior cerebellar artery. It penetrates the dura of the lateral side of the clivus about 1 cm below the crest of bone that continues from the side of the base of the dorsum sellae to the petrous ridge. It courses in the dura over this crest, in a slight groove. It passes under an extension of the attached margin of the tentorium cerebelli from the apex of the petrous temporal to the margin of the dorsum sellae and posterior clinoid process, known as the petrosphenoidal ligament (of Gruber), into the cavernous sinus in the middle cranial fossa. The passage into the cavernous sinus is known as Dorello's canal. It may be medial, inferior or lateral to the inferior petrosal sinus, or may even pass through it. Sometimes the nerve is double, when one nerve passes deep to the petrosphenoidal ligament and the other above it. The nerve is vulnerable to damage following concussion, fracture or infection of the petrous temporal, and to raised intracranial pressure, when a lateral rectus palsy results.

MASTOID AREA

423. Outline the suprameatal triangle.

The suprameatal, or Macewan's triangle, is bounded above by the supra-mastoid crest, below and in front by the posterosuperior margin of the external auditory meatus, and behind by a vertical tangent drawn down to the meatus from the crest. The meatal margin presents the suprameatal or Henle's spine. The triangle is deep to the cymba conchae, the hollow of the auricle between the root of the helix and the inferior crus of the antihelix. It overlies the mastoid antrum, forming its lateral wall, which can vary in thickness from 2–3 mm at birth up to 15 mm in the adult. The margins of the triangle are surface guides to the floor of the middle cranial fossa above, the descending part of the facial nerve in front and below, and the sigmoid sinus behind.

424. What is the cavity deep to the suprameatal triangle, and what are its relations ?

Deep to the suprameatal triangle is the mastoid antrum, an air sinus in the temporal bone. It has a volume of about 1 ml, its anteroposterior diameter being about twice its transverse diameter. It communicates through an irregular opening, the aditus, in the upper part of its anterior wall, with the epitympanic recess of the tympanic cavity. Lower down in front is the descending part of the facial nerve canal. Mastoid air cells lead off from the posterior, inferior, and sometimes anterior aspects of the antrum. Medial to the aditus, raising a rounded eminence above and behind the prominence of the facial nerve canal, is the lateral semicircular canal. medial to the medial wall of the antrum is the posterior semicircular canal. Behind and below the antrum, beyond the air cells, is the sigmoid sinus. Below and laterally is a ridge formed by the mastoid notch for the posterior belly of the digastric muscle. Above is the tegmen tympani and the temporal lobe of the cerebral hemisphere.

425. What are the features of the tympanic cavity ?

The tympanic cavity is the cavity in the petrous temporal bone containing the auditory ossicles, between the bony vestibule medially and the external auditory meatus laterally. It is biconcave in shape, and about 15 mm in anteroposterior and vertical diameters. Its width ranges from about 4 mm at the periphery to about 2 mm centrally. The **roof** of the cavity is formed by the tegmen tympani. The **floor** is formed by a plate of bone separating the cavity from the bulb of the internal jugular vein. The canaliculus for the tympanic branch of the glossopharyngeal nerve perforates the floor. The **anterior wall** is formed from below upwards by a thin plate of bone separating the cavity from the internal carotid artery and showing perforations for caroticotympanic sympathetic and vascular branches; the opening of the pharyngotympanic (auditory) tube; the opening for the canal for the tensor tympani, and the opening of the canaliculus for the lesser petrosal nerve passing forward to the middle cranial fossa. The **lateral wall** is formed below by the external auditory meatus, closed by the tympanic membrane, and above by the epitympanic recess, which curves slightly outwards above the medial end of the meatus. The posterior canaliculus for the chorda tympani opens at the junction of the lateral and posterior walls, and the anterior canaliculus opens at the junction of the lateral and anterior walls. The anterior canaliculus leads to an opening into the medial end of the petrotympanic suture. The **posterior wall** presents from above downwards the aditus to the antrum, the fossa incudis for the attachment of the short process of the incus, the conical pyramid through which projects the tendon of the stapedius muscle, and a thin plate of bone separating the cavity from the canal for the stapedius. The canal for the facial nerve runs downwards from medial to the aditus, behind the lower features of the posterior wall, to the stylomastoid foramen. The **medial wall** presents three swellings and two openings. The **three swellings** are a) the promontory, the most prominent bulge, formed by the first turn of the cochlea and minutely grooved by the tympanic plexus, b) the processus cochleariformis above the promontory, a hook-like projection formed by an extension of the plate of bone between the pharyngotympanic tube and the canal for the tensor tympani, and around which bends the tendon for the tensor tympani, and c) the ridge, close to the roof posteriorly, formed by the lateral semicircular canal. The **two openings** are a) the oval window (fenestra vestibuli) above and behind the promontory, closed by the footplate of the stapes, and b) the round window (fenestra cochlea) below and behind the promontory, closed by the secondary tympanic membrane. Between the promontory and the ridge of the lateral semicircular canal a thin plate of bone separates the tympanic cavity from the canal for the facial nerve, which runs backwards from opposite the epitympanic recess to the medial wall of the aditus. Behind the promontory is a small depression, the sinus tympani.

426. What does the mastoid process contain ?

The mastoid process contains air cells. These are interconnecting air-filled cavities which spread out from the mastoid antrum, to a varying degree. There may be a spread of air cells throughout the mastoid process, down to its apex. They may extend to either side of the facial canal, to lie against the tegmen, to adjoin the sigmoid sinus, and to partially surround the posterior semicircular canal. Superiorly and anteriorly a curtain of bone, called **Koerner's (petrosquamosal) septum**, may descend a short distance within the process partially dividing the system of air cells. It marks the plane of junction between the parts of the mastoid process derived from the squamous and petromastoid parts of the temporal bone. Air cells can extend anteriorly in the petrous part to lie close to the labyrinth, cochlea, carotid canal, and auditory canal, and may even reach the apex of the bone. If many air cells are present a mastoid process is described as **pneumatised**. Lesser development of air cells may leave marrow-containing cancellous bone between and around them, when the mastoid process is described as **diploic**. If there is dense bone between few air cells the process is described as **sclerotic**.

427. What are the features of the vestibule of the bony labyrinth ?

The vestibule is the central part of the bony labyrinth of the internal ear, medial to the tympanic cavity, lateral to the internal auditory meatus, anterior to the semicircular canals and posterior to the cochlea. Rather irregular in shape, it has an anteroposterior and vertical diameter of about 5 mm and a transverse maximum diameter of about 3 mm. Its **roof** is the tegmen. Its **floor** is above the posterior end of the carotid canal and the bulb of the internal jugular vein against the jugular fossa. The **anterior wall** leads through an elliptical opening into the scala vestibuli of the cochlea. **Posteriorly** are the five openings of the semicircular canals. The **lateral wall** has the fenestra vestibuli for the foot-plate of the stapes. The **medial wall** presents a number of features —

i. the **spherical recess**, anteriorly, for the saccule. It is perforated by the foramina of the macula cribrosa media, for branches of the vestibular nerve. It corresponds to the inferior vestibular area of the fundus of the internal auditory meatus.

ii. the **vestibular crest**, an oblique ridge behind the spherical recess. It forms a slight projection above, called the pyramid, which is perforated by foramina for branches of the vestibular nerve to the utricle. Below, the crest divides to outline a small depression, the cochlear recess, perforated by foramina for branches of the cochlear nerve.

iii. the **elliptical recess**, above and behind the spherical recess, for the utricle. It is perforated by foramina for branches of the vestibular nerve. The foramina of the pyramid and elliptical recess constitute the macular cribrosa superior. It corresponds to the superior vestibular area of the fundus of the internal auditory meatus.

iv. the **opening of the endolymphatic duct**, below the elliptical recess.

428. What nerves lie superficial to the mastoid area ?

Nerves that lie superficial to the mastoid process are —

the **posterior auricular nerve**, a branch of the facial nerve. It ascends between the mastoid process, and the external auditory meatus. It divides into an auricular branch which supplies the auricularis posterior and the cranial intrinsic muscles of the auricle, and an occipital branch which runs along the superior nuchal line to supply the occipital belly of the occipitofrontalis.

the **auricular branch of the vagus**. It emerges from the tympanomastoid suture and divides into a branch which joins the posterior auricular nerve, and another which supplies part of the cranial surface of the auricle, the posterior wall and floor of the external auditory meatus, and the adjoining part of the outer surface of the tympanic membrane.

the **posterior branch of the great auricular nerve (C2,3)**. It supplies the skin of the lower part of the cranial surface of the auricle, the concha and lobule of its outer surface, and the mastoid process.

the **lesser occipital nerve (C2)**. It ascends behind the great auricular nerve, connects with it, and supplies the upper part of the cranial surface of the auricle.

429. Define the external auditory meatus. What are its relations ?

It is the bony canal which leads from the exterior towards the tympanic cavity. It is closed medially by the tympanic membrane. Because of the oblique set of the membrane, which faces downwards, at an angle of about

55°, and slightly forwards as well as laterally, the canal is approximately 6 mm longer anteroinferiorly than posterosuperiorly. The average depth of the bony canal in the adult is about 15 mm, proportionately much smaller in the neonate. It passes inwards, and slightly forwards and downwards. The perimeter of the tympanic membrane is attached to the tympanic sulcus in the bone. The cartilaginous part of the external acoustic meatus is attached to the slightly roughened and dilated lateral end of the bony meatus. The anterior, inferior, and most of the posterior part of the meatus is provided by the tympanic plate of the temporal bone. An aperture, called the foramen of Huschke, may persist from fetal life in the floor of the meatus in the tympanic plate. The remaining superior part of the meatus is formed by the squamous part of the temporal bone. **Anterior** to the meatus is a thin layer of parotid gland, containing the auriculotemporal nerve and superficial temporal vessels, separating it from the temperomandibular joint. There may be a preauricular lymph node. **Posterior** to the meatus a thin plate of bone separates it from the mastoid air cells. **Superior** to the meatus is the middle cranial fossa. **Medially** a part of the epitympanic recess intervenes between it and the cranial cavity. **Inferior** to the meatus are, from lateral to medial, parotid gland, posterior belly of digastric, facial nerve, styloid process and, in tandem, the internal carotid artery anteriorly and the jugular bulb posteriorly. The deep auricular branch of the maxillary artery with its accompanying veins may penetrate the tympanic plate to supply the lining of the meatus and exterior of the tympanic membrane.

THE UNDERSURFACE OF THE CRANIUM

430. What bones make up the hard palate ?

The hard palate is made up of —
     the palatine processes of the **maxillae**.
     the horizontal plates of the **palatine** bones.
     the pyramidal processes (tubercles) of the palatine bones.

431. Identify the incisive fossa. What passes through it ?

The incisive fossa pierces the hard palate immediately posterior to the central incisor teeth. It overlies the incisive papilla of the palate, prominent in the child. It marks the position of the nasopalatine duct, the communication in the embryo between the oral and nasal cavities, which persists for a short time before being closed by the fusion of the palatine processes of the maxillae and the posterior ridge of the frontonasal process. Epithelial elements may remain to cause cysts in later life. Incisive canals from the nasal cavity open into the depths of the fossa through two incisive foramina, right and left, which give passage to the nasopalatine nerves to the palate, and vessels which communicate between the greater palatine and sphenopalatine vessels. Often there are four foramina, the additional two lying anteriorly and posteriorly, and transmitting the nasopalatine nerves separately from the vessels. It is often said that in these cases the left nerve passes through the anterior foramen and the right through the posterior foramen.

432. Identify the greater palatine foramen. What passes through it ?

The greater palatine foramen is the inferior opening of the greater palatine canal, leading down from the pterygopalatine fossa and emerging onto the palate behind the lateral limit of the palatomaxillary fissure, medial to the third molar tooth. The canal is formed by the union of opposing grooves on the maxilla and the perpendicular plate of the palatine bone. It transmits the greater palatine nerve from the pterygopalatine ganglion, and the greater palatine artery from the terminal part of the maxillary artery, and its accompanying veins.

433. Identify the lesser palatine foramen. What passes through it ?

The lesser palatine foramen is the lower opening of the lesser palatine canal in the pyramidal process of the palatine bone, which communicates above with the greater palatine canal. There are usually two foramina and canals. They transmit the lesser palatine branches of the greater palatine nerve and vessels.

434. Identify the palatine crest.

The palatine crest is a variably prominent ridge on the horizontal plate of the palatine bone which extends medially from just behind the greater palatine foramen. It marks the most anterior attachment of the tensor palati.

435. Identify the pharyngeal tubercle. What is it for?

The pharyngeal tubercle is the slight midline prominence on the under-surface of the basiocciput (clivus) about 1 cm in front of the foramen magnum, to which is attached the median raphe of the superior constrictor muscle of the pharynx.

436. Where is longus capitis inserted ?

Longus capitis is inserted into the prominence of the basilar part of the occipital bone anterolateral to the pharyngeal tubercle.

437. Identify the pterygoid fossa. What does it contain ?

The pterygoid fossa is the fossa between the medial and lateral pterygoid plates. It is occupied chiefly by the deep head of medial pterygoid, which arises from the medial surface of the lateral pterygoid plate, and the grooved posterior surface of the pyramidal process of the palatine bone at the lower end of the fossa. Tensor palati passes down to the pterygoid hamulus, medial to the medial pterygoid.

438. Identify the scaphoid fossa. What is it for ?

The scaphoid fossa is the elliptical groove formed by the division of the upper end of the posterior border of a medial pterygoid plate, directed towards the spine of the sphenoid and medial to the foramen ovale. The anterior part of the tensor palati arises from it.

439. Identify the pterygoid hamulus. What is attached to it ?

The pterygoid hamulus is the thin, hook-like process that curves downwards, outwards and backwards from the lower end of the medial pterygoid plate. The tendon of tensor palati turns acutely round it laterally and anteriorly in the deep notch at its base, to expand into the palatine aponeurosis. To its tip is attached the pterygomandibular raphe, and a short ligament (the pterygomaxillary ligament) to the pyramidal process of the palatine bone. A few fibres of buccinator arise from it anteriorly, and of superior constrictor posteriorly. The tip of the hamulus can be felt like a large pinhead in the root of the soft palate just behind and medial to the back of the upper alveolar process.

440. Identify the spine of the sphenoid. What is attached to it ?

The spine of the sphenoid is the prominent thin projection from the posterior angle of the greater wing of sphenoid. To its tip is attached the sphenomandibular ligament, which passes to the lingula of the mandible. Attached to its anterior border may be a pterygospinous ligament, from a part of the posterior border of the lateral pterygoid plate, which can be ossified. Fibres of the anterior malleolar ligament reach it posteriorly through the petrotympanic fissure. The spine forms the posterolateral margin of the groove for the cartilaginous part of the pharyngotympanic tube, which lies between it and the petrous part of the temporal bone. The lateral cartilaginous plate of the auditory tube lies against the upper part of the medial surface of the spine. From lower down the medial surface of the spine arise fibres of the tensor palati, and between these fibres and the spine, often in a groove in the bone, is the chorda tympani. In contact with the spine laterally is the lateral pterygoid muscle. Close to the spine are the middle meningeal vessels anteriorly, which pass through the foramen spinosum, and the internal carotid artery posteriorly. The auriculotemporal nerve passes backwards below and lateral to the tip of the spine, and splits around the middle meningeal artery.

441. Identify the groove for the cartilaginous part of the pharyngo-tympanic tube.

The groove for the cartilaginous part of the pharyngotympanic tube extends forwards and medially between the greater wing of the sphenoid and the petrous part of the temporal bone. The junction between these two bones is the petrosphenoidal suture. The roof of the groove may be deficient. Two canals lead back from the groove posteriorly to the interior of the temporal bone. The upper, smaller canal contains tensor tympani, which arises chiefly from the cartilage of the pharyngotympanic tube and passes back to insert into the malleus. The lower, larger canal is the bony part of the pharyngo-tympanic tube, and communicates with the tympanic cavity.

442. Identify the fissures between the tympanic plate and the mandibular fossa.

Laterally is the **squamotympanic fissure**. Medially this fissure is divided by a small protrusion of the petrous temporal bone which is the down-turned anterolateral edge of the tegmen tympani. The tegmen tympani is the thin plate of bone which projects laterally from the petrous part of the temporal bone to form, anteriorly, the roof of the canal for tensor tympani. The fissure anterior to this projection of bone is the **petrosquamous fissure**. The fissure posterior to the projection is the **petrotympanic fissure**, through which pass the chorda tympani and the anterior tympanic vessels. Fibres of the anterior malleolar ligament, a remnant of the perichon-drium of Meckel's cartilage, also traverse the petrotympanic fissure to the spine of the sphenoid.

443. Identify the inferior opening of the carotid canal. What passes through it ?

The inferior opening of the carotid canal is the prominent foramen just in front of the jugular foramen, separated from it by a narrow ridge of bone. Through it passes the internal carotid artery with its surrounding sympa-thetic and venous plexuses. Just within the canal a small foramen on its lateral wall transmits the caroticotympanic branch of the internal carotid artery with accompanying veins and sympathetic nerves to the tympanic cavity. The carotid canal passes forwards in the apex of the petrous part of the temporal bone to the foramen lacerum.

444. Identify the styloid process. What is attached to it ?

The styloid process projects downwards, forwards and medially from just lateral to the jugular foramen. It is one of the four developmental compo-nents of the temporal bone, partly sheathed at its base by the tympanic plate. It lies between the internal jugular vein and the parotid gland. It gives rise to two ligaments, the stylohyoid ligament from its tip to the lesser cornu of the hyoid bone, and the stylomandibular ligament, a thickening of the deep parotid fascia, from its anterior aspect to the posterior margin of the mandible. It gives rise to three muscles, the stylopharyngeus from its upper medial surface, the stylohyoid from the posterior surface of its middle third, and the styloglossus from the anterior surface near its tip. The facial nerve passes close to its lateral surface.

445. Identify the stylomastoid foramen. What passes through it?

The stylomastoid foramen lies closely posterolateral to the root of the styloid process. It is the inferior opening of the facial canal. From it emerges the facial nerve and accompanying veins. Into it passes the stylomastoid branch of either the posterior auricular or occipital arteries. It lies 2 cm deep to the middle of the anterior border of the mastoid process.

446. Identify the groove for the occipital artery on the mastoid part of the temporal bone.

The shallow groove, not always present, for the occipital artery, runs close to the anterior part of the squamo-occipital suture, medial to the mastoid notch.

447. Identify the mastoid notch. What is is for ?

The mastoid notch is the deep groove separating the upper medial part of the mastoid process from the remainder of the temporal bone, up to the squamo-occipital suture, lateral to the groove for the occipital artery. The posterior belly of digastric arises from the length of the groove.

448. What are the boundaries of the posterior nasal apertures ?

In the midline, apertures called **choanae** are separated by the posterior border of the vomer. Below, the vomer articulates with the posterior part of the nasal crest formed by the joined thick medial borders of the horizontal plates of the palatine bones. The inferior boundary of each aperture is the posterior border of the horizontal plate of a palatine bone. The lateral boundary is the posterior margin of a medial pterygoid plate. The superior boundary is formed by the ala of the vomer medially and the vaginal process of a pterygoid process laterally. The inferior surface of the vaginal process is notched by the pharyngeal branch of the pterygopalatine ganglion and accompanying branch of the maxillary artery. When the ala of the vomer tucks in above the vaginal process of the medial pterygoid plate, the posterior opening of the vomerovaginal canal, formed between them and containing small vessels, opens into the upper boundary of the aperture.

449. Where do the muscles that move the mandible arise from the cranium ?

Temporalis arises from the temporal fossa. The area is outlined above by the temporal line of the frontal bone, the inferior temporal line of the parietal bone and the supramastoid crest of the temporal bone. The origin extends down to the contiguous infratemporal crests of the squamous part of the temporal bone and greater wing of the sphenoid. It includes very little of the upper surface of the posterior part of the zygomatic process of the temporal bone, or the posterior surface of the zygomatic bone, over which the muscle fibres play.

Masseter has superficial and deep parts. The superficial part arises by strong tendinous fibres from the lower border of the zygomatic bone. Anteriorly the origin may spread to the outer corner of the zygomatic process of the maxilla. Posteriorly tendinous fibres do not arise beyond the zygomaticotemporal suture. The deep part arises by muscle fibres from the inner surface of the zygomatic arch. Anteriorly it is overlapped by the superficial tendinous part. Posteriorly the origin extends to the lower border of the zygomatic process of the temporal bone.

Lateral pterygoid arises by two heads. The upper head arises from the infratemporal surface and crest of the greater wing of the sphenoid. The lower head arises from the lateral surface of the lateral pterygoid plate.

Medial pterygoid arises also by two heads. The deep head arises from the medial side of the lateral pterygoid plate, the pterygoid fossa and the posterior surface of the pyramidal process of the palatine bone. The superficial head arises from the tuberosity of the maxilla and the lateral surface of the pyramidal process of the palatine bone.

The posterior belly of **digastric** arises from the length of the mastoid notch of the temporal bone.

450. How does the pharynx attach to the cranium ?

Superior constrictor arises on each side from the lower end of the posterior margin of the medial pterygoid plate and the pterygoid hamulus. The **medial pharyngeal raphe** inserts into the pharyngeal tubercle of the basilar part of the occipital bone. Between these two attachments on each side, the **pharyngobasilar fascia** is attached to the basilar part of the occipital bone anterior to the attachment of longus capitis, to the cartilage filled petro-occipital suture, to the petrous part of the temporal bone anterior to the carotid foramen, to the undersurface of the cartilaginous part of the auditory tube, and to the upper part of the posterior border of the medial pterygoid plate. **Stylopharyngeus** arises from the medial surface of the upper part of the styloid process of the temporal bone. **Salpingopharyngeus** arises from the undersurface of the cartilaginous part of the pharyngotympanic tube near its opening into the pharynx. Middle constrictor arises from the lower part of the stylohyoid ligament and the upper border of the greater cornu of the hyoid, and is thus suspended from the cranium.

451. Where are the muscles of the palate attached to the cranium?

Tensor palati arises from the scaphoid fossa of the pterygoid process of the sphenoid bone, the lateral cartilaginous lamina of the pharyngotympanic tube, and the medial surface of the spine of the sphenoid (i.e. outside the pharynx).

Levator palati arises from the inferior surface of the petrous part of the temporal bone anterior to the carotid foramen, and from the inferior surface of the cartilaginous part of the pharyngotympanic tube (i.e. inside the pharynx).

Anterior fibres of **palatopharyngeus** arise from the posterior border of the hard palate.

Some fibres of **musculus uvulae** arise from the posterior nasal spine formed by the palatine bones.

452. Where are the muscles of the neck attached to the cranium ?

Longus capitis is inserted into the basilar part of the occipital bone, anterolateral to the pharyngeal tubercle.

Rectus capitis anterior is inserted into the basilar part of the occipital bone, posterior and lateral to longus capitis, in front of the occipital condyle.

Rectus capitis lateralis is inserted into the rough inferior surface of the jugular process of the occipital bone.

Rectus capitis posterior minor is inserted into the medial part of the area of the squamous part of the occipital bone, between the foramen magnum and the inferior nuchal line.

Rectus capitis posterior major is inserted into the lateral part of this area.

Semispinalis capitis is inserted into the medial part of the area of the squamous part of the occipital bone between the inferior and superior nuchal lines.

Obliquus capitis superior is inserted into the lateral part of this area.

Longissimus capitis is inserted into the posterior margin of the mastoid process of the temporal bone, deep to splenius capitis.

Splenius capitis is inserted into the mastoid process anterior to longissimus capitis, and to the squamous part of the occipital bone, below the lateral third of the superior nuchal line and above part of semispinalis capitis.

Sternocleidomastoid is inserted by a strong tendon into the apex and lateral surface of the mastoid process, and by a thin aponeurosis into the lateral half of the superior nuchal line.

Trapezius is inserted by aponeurotic fibres into the external occipital protuberance and the medial half of the superior nuchal line.

453. What is the infratemporal fossa ?

The infratemporal fossa is the region below the lateral part of the middle cranial fossa. The roof is the infratemporal area of the cranium formed mostly by the greater wing of the sphenoid in front of a small area of the squamous part of the temporal, and bounded laterally by the infratemporal crest. Anteriorly is the posterior wall of the maxilla, below the inferior orbital fissure, and lateral to the pterygomaxillary fissure. Medially is the lateral pterygoid plate. Laterally is the ramus of the mandible. Posteriorly is the styloid process. Inferiorly the fossa continues into the neck. The fossa contains the pterygoid muscles, the maxillary artery and its branches, the pterygoid plexus of veins, branches of the mandibular nerve, chorda tympani and the otic ganglion. The posterior wall of the fossa is completed by the muscles arising from the styloid process, and the carotid sheath.

THE ORBIT

454. What bones make up the margin of the orbit ?

The superior orbital margin is formed by the **frontal** bone, with the supraorbital notch or foramen lying at the junction of the round medial third with the sharp lateral two thirds. The lateral orbital margin, the strongest, is formed mostly by the frontal process of the **zygomatic** bone, but with a small part of the zygomatic process of the frontal bone above it. The inferior orbital margin is formed about equally by the maxillary process of the zygomatic bone and the **maxilla**. The medial orbital margin is not continuous. The inferior margin of the orbit passes up anteriorly into the lacrimal crest of the frontal process of the maxilla. The superior margin of the orbit passes down posteriorly into the crest of the **lacrimal** bone. The overlapping crests bound the groove for the lacrimal sac, which leads into the nasolacrimal canal.

455. What bones make up the walls of the orbit ?

Seven bones contribute to the walls of the orbit; frontal, sphenoid, maxilla, zygomatic, palatine, lacrimal and ethmoid.

The **superior** wall, or roof, is formed mostly by the orbital plate of the frontal bone, but posteriorly the lesser wing of the sphenoid contributes a small part above the superior orbital fissure.

The **inferior** wall, or floor, is formed mostly by the orbital plate of the maxilla, but also by the orbital surface of the zygomatic bone anterolaterally, and by the orbital process of the palatine bone posteriorly.

The **lateral** wall is formed by the orbital surface of the frontal process of the zygomatic bone anteriorly and the orbital surface of the greater wing of the sphenoid posteriorly.

The **medial** wall is formed from front to back by the frontal process of the maxilla, the lacrimal bone, the orbital plate of the ethmoid and the body of the sphenoid.

456. What are the important relations of the orbit ?

Superiorly the frontal sinus can extend backwards in the medial part of the roof of the orbit to a varying degree. The frontal lobe of the cerebral hemisphere lies in the anterior cranial fossa above the orbit. **Medial** to the anterior part of the medial wall of the orbit, and medial to the lacrimal groove containing the lacrimal sac, is the nasal cavity. Behind the lacrimal groove the orbit is related medially to the ethmoidal and sphenoidal air sinuses. The orbital plate of the labyrinth of the ethmoid bone is very thin, and is known as the lamina papyracea. Spread of infection from an ethmoidal sinus is said to be the most common cause of orbital cellulitis. **Inferiorly** is the maxillary sinus. The canal of the infraorbital nerve is within the floor of the anterior part of the orbit. This wall is thin, especially posteromedially, where it is subject to fracture in orbital blow-out injuries. **Laterally** and **posteriorly** is the middle cranial fossa and the temporal lobe. This is the weakest portion of the lateral wall. **Laterally** and **anteriorly** is the temporal fossa, a route of surgical access to the contents of the orbit.

457. How is the eyeball related to the bony orbit ?

The eyeball, about 6.5 ml in volume, lies anteriorly within the orbit, which has a capacity of about 30 ml. It is placed nearer the roof than the floor, and is slightly nearer the lateral wall than the medial wall. The cornea does not normally protrude beyond a line joining the superior and inferior margins of the orbit, but nearly a third of the eyeball protrudes beyond a line joining the medial and lateral margins.

458. Identify the superior orbital fissure. What passes through it ?

It is the retort-shaped fissure between the posterior parts of the roof and lateral wall of the orbit, between the greater and lesser wings of the sphenoid bone, for communication between the middle cranial fossa and the orbit. It is divided into lateral, intermediate and medial parts by encroachment across it of the common anular tendon which gives origin to the four rectus muscles to the eye. The **lateral part** of the fissure is the narrowest and is outside the common anular tendon. It gives passage to the lacrimal nerve, the frontal nerve and the trochlear nerve, in that order from lateral to medial, in a group just lateral to the origin of the lateral rectus. The outermost part of the fissure may contain a communicating artery between the anterior branch of the middle meningeal artery within the middle cranial fossa and the lacrimal branch of the ophthalmic artery within the orbit. This branch, however, often passes through its own foramen, visible on X-ray films of the orbit, just in front and lateral to the lateral limit of the fissure, called by radiologists Hyrtl's foramen. A superior ophthalmic vein may also pass through the lateral part of the fissure to the cavernous sinus. The **intermediate part** of the fissure, sometimes known as the oculomotor foramen, is bounded by the common anular tendon, and through it pass from above downwards the superior branch of the oculomotor nerve, the nasociliary nerve and the inferior branch of the oculomotor nerve. In its course through the foramen the position of the abducent nerve moves from inferior to the inferior branch of the oculomotor nerve to a position lateral to the nasociliary nerve. The sympathetic and sensory roots of the ciliary ganglion accompany the nasociliary nerve through this part of the fissure, either separately or incorporated in the nerve. An ophthalmic vein may pass through the intermediate part of the fissure to the cavernous sinus. The **medial part** of the fissure is the widest, and through it may pass an inferior ophthalmic vein to the cavernous sinus.

459. Identify the inferior orbital fissure. What passes through it ?

It is the fissure between about the posterior three quarters of the floor and lateral wall of the orbit, bounded by the greater wing of the sphenoid above, the zygomatic bone anteriorly, and the maxilla and the orbital process of the palatine bone below. The zygomatic bone may be excluded by fusion of the greater wing with the maxilla at the anterior limit of the fissure. The fissure provides communication between the orbit and the pterygopalatine fossa posteriorly and between the orbit and the infratemporal fossa anteriorly.

Passing through the fissure are —

i. the maxillary nerve. It then becomes the infraorbital nerve, which courses anteriorly and medially at first in the infraorbital groove and then through the infraorbital canal.

ii. vessels which accompany the infraorbital nerve.

iii. the zygomatic branch of the maxillary nerve, which passes forwards and laterally to divide into its zygomaticofacial and zygomaticotemporal branches which pass through separate canals in the zygomatic bone.

iv. branches of the pterygopalatine ganglion. These pass to the periosteum of the orbit (called the periorbita), the orbitalis muscle of Muller, overlying the fissure, and through the posterior ethmoidal foramen to the ethmoidal and sphenoidal sinuses. Other branches join with branches from the internal carotid plexus to form the retro-orbital plexus of nerves, which supplies other orbital structures, probably including the lacrimal gland.

460. Identify the anterior ethmoidal foramen. What passes through it ?

The anterior ethmoidal foramen is found in, or just above, the fissure between the orbital plate of the ethmoid and the orbital plate of the frontal bone, about 1.5 cm behind the suture with the lacrimal bone. The foramen is the proximal opening of the anterior ethmoidal canal, which courses anteromedially in the roof of the ethmoidal sinuses, to emerge at the lateral margin of the superior surface of the cribriform plate. It transmits the anterior ethmoidal nerve and vessels, which supply the sinuses and pass on to supply the nasal cavity. The anterior ethmoidal nerve becomes the external nasal.

461. Identify the posterior ethmoidal foramen. What passes through it ?

The posterior ethmoidal foramen is found in, or just above, the fissure between the orbital plate of the ethmoid and the orbital plate of the frontal bone, about 1 cm in front of the optic foramen. It is the orbital opening of the posterior ethmoidal canal which courses medially in the roof of the ethmoidal and sphenoidal sinuses to emerge onto the superior surface of the posterolateral corner of the cribriform plate, between it and the body of the sphenoid bone. It transmits the posterior ethmoidal vessels. The posterior ethmoidal nerve is often absent and is said not to reach the dura.

462. Identify the intraorbital foramina for the zygomatic nerve or its branches.

Small furrows may lead from the anterior end of the inferior orbital fissure to two foramina on the orbital surface of the zygomatic bone. The upper foramen leads into a canal which transmits the zygomaticotemporal nerve to the temporal fossa and the lower foramen leads into a canal which transmits the zygomaticofacial nerve onto the cheek. The zygomatic nerve may enter the zygomatic bone as a single nerve, branching in the bone.

463. Identify the infraorbital groove. What is it for ?

The infraorbital groove runs forwards and medially in the floor of the orbit from the anterior margin of the inferior orbital fissure to the posterior opening of the infraorbital canal. It accommodates the infraorbital nerve and vessels. The groove and canal project into the roof of the maxillary sinus. Bony defects may expose the infraorbital nerve in parts to the mucous membrane of the sinus. Within the groove and canal the infraorbital nerve gives off the middle and anterior superior alveolar nerves.

464. Where are the extraocular muscles of the eye attached to the walls of the orbit ?

The **four rectus muscles** arise in continuity from the **common anular tendon** (of Zinn), a fibrous thickening of the periorbita which extends obliquely in an elliptical manner around the anterior opening of the optic canal, across the more medial part of the superior orbital fissure, to a small tubercle on the posterior margin of the greater wing of the sphenoid. Around the superior and medial aspects of the anterior opening of the optic canal the common anular tendon occupies the angle of divergence of the dural sheath of the optic nerve and the periorbita. The upper part of the common anular tendon which gives rise to the upper part of the lateral rectus, the superior rectus, and the upper part of the medial rectus muscles is said to constitute the **tendon of Lockwood**. The part of the common anular tendon which arises from the inferior root of the lesser wing of the sphenoid, often called the optic strut, is specialised to the extent that it receives the name of the **tendon of Zinn**. Its attachment is sometimes marked by an infraoptic tubercle. It gives origin to the lower part of the medial rectus, the inferior rectus, and part of the lower part of the lateral rectus muscles.

Levator palpebrae arises by a narrow tendon from the lesser wing of the sphenoid above, but blending with, the origin of the superior rectus muscle.

Superior oblique arises by a narrow tendon from the body of the sphenoid, above and medial to the optic canal, outside the common anular tendon, near the frontoethmoid suture. It partially overlaps the origin of levator palpebrae. The tendon of superior oblique passes through and around the saddle-shaped cartilaginous trochlea. This is attached by dense fibrous tissue to a small circular dimple in the wall of the orbit about 4–5 mm behind the orbital margin, at the superior nasal angle of the orbit. The site of attachment of the trochlea may be a small bony tubercle rather than a fovea.

Inferior oblique arises by a narrow tendon from the orbital surface of the maxilla just lateral to the margin of the opening of the nasolacrimal canal, and near to the rim of the orbit.

465. Where is orbicularis oculi attached ?

Three parts to the muscle are described —

the **palpebral part**, confined to the eyelids. It arises medially from the medial palpebral ligament, a flattened band connecting the medial ends of the tarsal plates to the frontal process of the maxilla, between its lacrimal crest and its suture line with the nasal bone. The fibres of the palpebral muscle pass laterally to the lateral palpebral raphe, which does not attach to bone.

the **orbital part**, which arises from the rim of the orbit medial to the supraorbital notch, round to the insertion of the medial palpebral ligament, and on to near the infraorbital foramen.

the **lacrimal part**, a thin layer of fibres which arises from the upper part of the crest of the lacrimal bone, behind the lacrimal sac.

466. Where are the suspensory mechanisms of the eye attached within the orbit ?

i. the **fascia bulbi (Tenon's capsule)** below the eye, between the inferior rectus above and the inferior oblique below, where they cross each other, is thickened to form the **suspensory ligament of Lockwood**. The ligament continues medially from the medial rectus as the medial check ligament, and is inserted, together with the tip of the medial horn of the tendon of the levator palpebrae, into bone just behind the crest of the lacrimal bone. The ligament continues laterally from the lateral rectus as the lateral check ligament, and is inserted, together with the tip of the lateral horn of the levator palpebrae and the lateral palpebral ligament, into the orbital tubercle (also called the marginal tubercle or Whitnall's tubercle), just inside the orbital margin, about 1 cm below the suture between the zygomatic process of the frontal bone and the frontal process of the zygomatic bone. Displacement of the zygomatic bone therefore causes double vision, but loss of the maxilla and the zygomatic bone below the tubercle may not.

ii. **Septa of orbital connective tissue** extend from the fascia bulbi to the periorbita, and form a framework which is thought to play a role in preventing downward displacement of the eye.

467. Where is the lacrimal apparatus in respect to the bone of the orbit ?

The **orbital lobe of the lacrimal gland** is lodged in a shallow fossa in the anterolateral part of the roof of the orbit behind the zygomatic process of the frontal bone. Loose connective tissue trabeculae pass from the stroma of the gland to the periorbita of the fossa. They sometimes cause small pits in the bone, and are said to constitute the suspensory ligament of the gland. The **lacrimal sac** lies in the groove between the crest of the lacrimal bone and the lacrimal crest of the frontal process of the maxilla. The groove is roofed over by periorbita, which splits around the sac to form the lacrimal fascia. The sac receives the lacrimal canaliculi and drains into the inferior meatus of the nasal cavity via the nasolacrimal duct, which passes down the nasolacrimal canal.

THE FACE

468. How is the scalp attached to the cranium ?

The foundation of the scalp is the occipitofrontalis muscle and its galea aponeurotica or epicranial aponeurosis. Posteriorly each belly of the occipitalis is a broad, thin, flat sheet attached by short tendinous fibres to the lateral two thirds of the highest nuchal line of the occipital bone, and more laterally to the adjacent part of the mastoid part of the temporal bone. Between the occipitalis bellies the galea is attached directly to the inner thirds of the highest nuchal lines and to the external occipital protuberance. Laterally, on each side, the galea thins to form a layer which passes over the temporal fascia and temporalis to attach to the zygomatic arch. Anteriorly the frontalis bellies merge in the midline. They do not insert into bone, but mingle with the procerus, corrugator supercilii, and orbicularis oculi muscles, and reach the skin of the forehead and eyebrows. The galea splits anteriorly to enclose the frontalis bellies. The superficial layer is thin, but the deep layer is better defined and is firmly attached on each side to the supra-orbital margin, contributing to the arcus marginalis, a thickening of the periorbita (the periosteum of the orbit) to which the orbital septum is attached. Between the occipitofrontalis and the underlying periosteum there is only loose connective tissue.

469. Where do the muscles of facial expression arise from the cranium ?

Procerus arises from the lower part of the nasal bone and the upper lateral nasal cartilage.

Corrugator supercilii arises from the frontal bone below the medial end of the superciliary arch.

Compressor naris arises from the maxilla lateral to the lower part of the nasal aperture.

Dilator naris arises from the maxilla below the lateral part of the lower margin of the nasal aperture.

Depressor septi arises from the maxilla above the central incisor tooth.

Levator labii superioris alaeque nasi arises from the upper part of the frontal process of the maxilla.

Levator labii superioris arises from above the infraorbital foramen, across the suture between the maxilla and zygomatic bone.

Levator anguli oris arises from the maxilla, from the canine fossa below the infraorbital foramen.

Zygomaticus minor arises from the zygomatic bone close to the zygomaticomaxillary suture, anterior to the zygomaticofacial foramen.

Zygomaticus major arises from the zygomatic bone close to the zygomaticotemporal suture, posterior to the zygomaticofacial foramen.

Orbicularis oculi is described in question 465.

The upper part of **Buccinator** arises from the labial aspect of the alveolar process of the maxilla, above the molar teeth, from the tuberosity of the maxilla and the tubercle of the palatine bone, and from a connecting band to the pterygoid hamulus. (See question 483 for its lower origin).

470. Where do the sensory nerves to the face emerge from the skull ?

Ten sensory nerves emerge from the skull to supply the face —

The **lacrimal** nerve, from below the lateral part of the rim of the orbit.

The **supraorbital** nerve, through the supraorbital notch or foramen.

The **supratrochlear** nerve, from within the rim of the orbit between the supraorbital foramen and trochlea. It may occupy a shallow furrow in the bone, the frontal notch, or else a frontal foramen.

The **infratrochlear** nerve, from within the rim of the orbit below the trochlea, but above the medial palpebral ligament.

The **external nasal** nerve, from below the lower border of the nasal bone, between it and the lateral nasal cartilage.

The **infraorbital** nerve, through the infraorbital foramen of the maxilla, which lies 0.5–1 cm below the middle of the inferior rim of the orbit, below the slight roughness palpable at the zygomatico-maxillary suture, and about 0.5 cm medial to a vertical line dropped from the centre of the pupil.

The **zygomaticofacial** nerve, through the zygomaticofacial foramen on the lateral surface of the zygomatic bone near its orbital border.

The **zygomaticotemporal** nerve, through the zygomaticotemporal foramen on the temporal surface of the zygomatic bone near the base of its frontal process.

The **mental** nerve, through the mental foramen, which usually lies halfway between the upper and lower borders of the body of the mandible, between the two premolar teeth. It may be a little lower, or under the second premolar. In the newborn the foramen opens anteriorly, but during the first 2 years of life this changes to its final posterosuperior direction.

The buccal and auriculotemporal nerves are branches of the **mandibular** nerve which emerges through the foramen ovale. The foramen ovale can be reached by a needle introduced horizontally in the coronal plane below the smooth inferior border of the posterior part of the zygomatic arch, just in front of the articular tubercle (eminence) at a depth of about 3.5–4 cm.

471. What are the boundaries of the anterior nasal aperture and what can be seen through it ?

The anterior nasal aperture is bounded on each side above by the inferior border of the nasal bone, and laterally and below by the nasal notch of the maxilla. A projection of the lower medial end of the nasal notch forms half of the anterior nasal spine.

Looking directly backwards into the nasal cavity it is seen to be divided into two by a septum which is commonly deviated from the midline. There is a posteriorly directed V-shaped defect in the bony septum for the septal cartilage. The upper part of the bony septum is formed by the perpendicular plate of the ethmoid bone, the lower part by the vomer and the nasal crests of the maxillae and palatine bones. The presenting edge of the vomer is grooved to receive the cartilaginous septum. Only the inferior and middle conchae can be seen. On each side, beneath the inferior concha is the inferior meatus. Beneath the middle concha is the middle meatus. Above the middle concha is the superior meatus. The superior concha is set well back out of sight on the lateral wall of the nasal cavity.

Looking upwards the cribriform plate of the ethmoid can be seen in the roof of the nasal cavity. The anterior border of the perpendicular plate of the ethmoid can be seen to articulate with the crests of the nasal bones. A groove on the undersurface of each nasal bone is occupied by the anterior ethmoidal nerve.

Looking downwards the floor of the nasal cavity is seen to be formed on each side by the palatine process of the maxilla anteriorly and the horizontal plate of the palatine bone posteriorly. The upper ends of the incisive canals can be seen anteriorly near the midline.

Looking laterally the inferior concha prevents a view of the opening of the nasolacrimal canal into the inferior meatus, but it may be possible to see through the maxillary hiatus into the back of the maxillary sinus, under and lateral to the middle concha, if the septum is deviated to the other side.

472. From knowledge of the local anatomy what signs might be expected to follow a fracture displacement of a zygomatic bone ?

Traumatic displacement of a zygomatic bone is usually inwards, backwards and more often downwards than upwards, resulting in loss of prominence of the cheek, though this can soon be disguised by the swelling which follows. Periorbital and subconjunctival haemorrhage may soon become evident. The subconjunctival haemorrhage has no visible posterior limit since the bleeding is occurring behind the conjunctival reflexion. Displacement of the zygomatic bone involves fractures near its sutures with neighbouring bones, namely with the maxilla in the floor of the orbit, with the greater wing of the sphenoid in the lateral wall of the orbit, with the zygomatic process of the temporal in the zygomatic arch, and with the frontal bone at the zygomaticofrontal suture, at the superolateral corner of the orbital margin. **Damage to the orbit** can lead to blurring of vision, usually temporary, from a concussive effect on the eyeball. Diplopia from displacement of the attachment of the lateral part of the suspensory ligament of Lockwood to Whitnall's tubercle just below the zygomaticofrontal suture will occur. The pupil of the affected side may be demonstrably at a lower level than that of the unaffected eye. muscle, most often the inferior oblique, or orbital fat may be damaged or trapped in a fracture line thereby causing restriction of movement of the eye. Orbital fat contains radial septa of connective tissue which render it particularly liable to entrapment. Thus there may be difficulty particularly in looking upwards and outwards, and diplopia on attempting to do so. There may be enophthalmos from herniation of fat into the maxillary sinus. **Damage to the maxilla** can lead to disruption of its sinus. Blood overflowing from the sinus causes epistaxis from the nostril of the same side. Since the bleeding arises in the sinus rather than from the nasal cavity, it cannot be controlled by packing. Blood in the sinus shows as an opaque shadow on an X-ray film. Escape of air can cause surgical emphysema of the orbit and cheek, and can be much increased, if the nose is blown. Palpation of the infraorbital margin may demonstrate a fracture step. Damage to the infraorbital, zygomatic, and anterior, middle and superior alveolar nerves will cause appropriate loss of sensation, usually temporarily. There will be swelling, and perhaps bruising of the upper gum and mucosa of the cheek. **Damage to the zygomatic arch** can lead to difficulty in opening the mouth, resulting from impaction of the coronoid process of the mandible against the depressed bone, or by bleeding or swelling, causing pain, in the masseter and temporalis muscles.

473. Identify the pterygomaxillary fissure. To where does it lead ?

The pterygomaxillary fissure is the thin triangular opening leading to the pterygopalatine fossa. It is bounded behind by the root of the pterygoid process of the sphenoid, and in front by the posterior wall of the maxilla. Below, at the apex of the triangle, these boundaries meet. Above, at the base of the triangle, they are separated by the posterior end of the inferior orbital fissure, below the lower border of the greater wing of the sphenoid. The pterygomaxillary fissure gives passage to the third part of the maxillary artery into the pterygopalatine fossa. Emerging from the fossa are veins to the pterygoid plexus, and the posterior superior alveolar nerve from the maxillary nerve.

474. What are the boundaries of the pterygopalatine fossa ?

The pterygopalatine fossa is roughly wedge-shaped. The base of the wedge is uppermost. The sharp edge points downwards. A **lateral wall** is missing; its place is taken by the pterygomaxillary fissure. The **medial wall** is the perpendicular plate of the palatine bone. There is an opening in the upper part of the medial wall, the sphenopalatine foramen, which is bounded by the orbital and sphenoidal processes of the perpendicular plate, and the body of the sphenoid. The **posterior wall** is the anterior surface of the root of the pterygoid process and adjoining part of the greater wing of the sphenoid. It has three foramina. The largest and most lateral is the anterior opening of the foramen rotundum. Medial to the foramen rotundum is the opening of the pterygoid canal. The smallest opening is most medial; it is the anterior opening of the palatovaginal canal (see question 448). The **anterior wall** is formed by the upper medial part of the posterior surface of the maxilla, with a small contribution from the orbital process of the palatine bone. The **roof** is missing. It is the posterior part of the inferior orbital fissure leading into the orbit. The **edge of the wedge** is formed below by the apposition of the maxilla, perpendicular plate of the palatine and pterygoid process of the sphenoid. There is a foramen in the edge leading into the palatine canal, formed between the maxilla and perpendicular plate of the palatine bone.

475. What does the pterygopalatine fossa contain ?

The pterygopalatine fossa contains —
    i. The **maxillary nerve**, the intermediate division of the trigeminal nerve, and its branches. It enters the fossa posteriorly through the foramen rotundum. It passes forwards and laterally across the upper part of the fossa, over the back of the orbital process of the palatine bone and the upper part of the back of the maxilla, through the posterior part of the inferior orbital fissure into the infraorbital groove where, thereafter, it is called the infra-orbital nerve. Within the fossa it gives off the zygomatic nerve, through the inferior orbital fissure, and the posterior superior alveolar nerve, through the pterygomaxillary fissure.
    ii. The **pterygopalatine ganglion** and its roots and branches. It is the largest of the parasympathetic ganglia. It lies in front of the anterior opening of the pterygoid canal, lateral to the sphenopalatine foramen and below and medial to the maxillary nerve, to which it is connected by a single trunk

more commonly than by separate branches. Preganglionic fibres reach the ganglion from behind, from the facial nerve, via the greater petrosal nerve and nerve of the pterygoid canal. It distributes postganglionic axons to the lacrimal gland, and the glands of the nasal, palatine and nasopharyngeal mucosa. Postganglionic fibres pass to the lacrimal gland by two possible routes. They are usually said to pass in its connection to the maxillary nerve, and then by the zygomatic, zygomaticotemporal and lacrimal nerves. Or they may pass from the ganglion in branches through the inferior orbital fissure to the retro-orbital plexus, which in turn supplies branches to the gland. Postganglionic fibres to the nose pass in the posterior superior nasal and nasopalatine branches through the sphenopalatine foramen. Fibres to the palate pass in the greater and lesser palatine branches through their palatine foramina. Fibres to the nasopharynx pass in the pharyngeal branch through the palatovaginal canal. The branches of the pterygopalatine ganglion also carry postganglionic sympathetic fibres, the cell bodies of which are in the superior cervical sympathetic ganglion. They reach the pterygopalatine ganglion in the internal carotid plexus, deep petrosal nerve, and the nerve of the pterygoid canal. The branches of the pterygopalatine ganglion also carry sensory fibres. Those of general sensation leave the pterygopalatine ganglion in its connection to the maxillary nerve, to join their cell bodies in the trigeminal ganglion. Those of taste, from the palate, leave the pterygopalatine ganglion in the nerve of the pterygoid canal and greater petrosal nerve, to join their cell bodies in the ganglion of the facial nerve.

iii. **The third part of the maxillary artery and its accompanying veins.** It enters the fossa between the two heads of the lateral pterygoid muscle, through the pterygomaxillary fissure. Passing medially in front of the pterygopalatine ganglion it terminates as the sphenopalatine artery which passes on through the sphenopalatine foramen. Branches within the fossa accompany the nerves through the infraorbital, palatine, pterygoid and palatovaginal canals.

476. What are the features of the nasolacrimal canal ?

The nasolacrimal canal is the passage, just over 1 cm long, which continues the lower end of the lacrimal groove behind the medial rim of the orbit, to the inferior meatus of the nasal cavity. Almost vertical, it slopes a little laterally and backwards. It conducts the nasolacrimal duct from the nasolacrimal sac, which lies in the lacrimal groove. The canal is formed by apposition of the lacrimal groove of the maxilla, with the lacrimal groove of the lacrimal bone and the lacrimal process of the inferior concha. The lacrimal groove of the maxilla is wide and passes down in front of the opening of the maxillary sinus. The lacrimal groove of the lacrimal bone is bounded behind by the lacrimal crest, which sweeps round the posterolateral edge of the upper opening of the canal as the lacrimal hamulus. The lower part of the floor of the groove of the lacrimal bone is prolonged downwards as its descending process, to join the lacrimal process of the inferior concha, which completes the lower opening of the canal, at the apex of the meatus, at the junction of the anterior and middle thirds. The nasolacrimal canal forms a ridge which projects a little into the maxillary sinus. It can be involved in fractures of the middle third of the face and can be invaded by a growth within the sinus, blocking the duct and causing an overflow of tears.

477. What nerves pass across the zygomatic arch ?

Passing upwards across the posterior root of the zygoma is the **auriculo-temporal nerve**, behind the superficial temporal vessels. More importantly, anteriorly, pass upwards the temporal branches of the **facial nerve**. There are usually four branches. They are evenly distributed in a 2.5 cm zone across the middle of the zygomatic arch. They supply the frontalis, corrugator supercilii and upper orbicularis oculi. The supply to the frontalis is usually by a single branch. The clinical effects of division of the nerves are loss of creases of the forehead, a lowering of the eyebrow and an inability to raise the eyebrow. There is no loss of ability to close the eyelid.

478. What is the nerve supply of the teeth of the upper jaw ?

The upper teeth are supplied by the anterior, middle and posterior superior alveolar nerves. Branches of these nerves within the alveolus join together to form a loose plexus above the roots of the teeth. From this plexus further branches arise which pass through root canaliculi to the pulp cavities and deep dentine of the teeth, into interradicular septa to supply the periodontal membrane of a single tooth, and into interalveolar septa to supply the periodontal membranes of adjacent teeth and the overlying gum. They are the only supply of the buccal gum, except posteriorly.

The **posterior superior alveolar nerve** is a branch of the maxillary nerve within the pterygopalatine fossa. It passes downwards and forwards through the pterygomaxillary fissure, to the posterior surface of the maxilla, which it penetrates as one to three branches. It usually gives a branch to the posterior part of the buccal surface of the gum. Within the maxilla the nerves pass in small canals in the posterior wall of the maxillary sinus. They supply the molar teeth.

The **middle superior alveolar nerve** is a branch of the infraorbital nerve, the continuation of the maxillary nerve in the orbit. It arises in the posterior part of the infraorbital canal. It is absent, however, in 60% of subjects. In others it may be duplicated. It passes laterally and downwards in the roof and side wall of the maxillary sinus and supplies the premolar teeth.

The **anterior superior alveolar nerve** is a branch of the infraorbital nerve which arises in the middle of the infraorbital canal. It passes in a sinuous canal in the wall of the maxillary sinus, laterally, forwards, downwards and then medially around the infraorbital foramen, to end in branches below the anterior nasal cavity. It supplies the canine and incisor teeth.

The **lingual gum** receives additional branches from the greater palatine and nasopalatine nerves. Branches of the nasopalatine nerve reach to the contralateral side near the midline, which explains why anaesthetic block of ipsilateral alveolar nerves is not always effective in the region of the ipsilateral central incisor.

The **buccal gum** may receive a few branches from the buccal nerve posteriorly.

The alveolar walls of the upper jaw are porous enough to allow local infiltration of the gum to anaesthetise the underlying teeth effectively enough for most dental procedures. Complete anaesthesia, to allow more severe procedures to be carried out on the upper jaw, requires the greater palatine and nasopalatine nerves to be blocked. The posterior superior alveolar nerves can be reached by a needle behind the maxilla. The anterior superior alveolar nerve can be reached through the infraorbital foramen, but the middle superior alveolar nerve is virtually inaccessible.

THE MANDIBLE

479. How does the mandible articulate with the cranium ?

The mandible articulates with the cranium, on each side, by a temporo-mandibular joint, which has special features —

i. the **shape of the articular surfaces** are unique. The cranial articular surface is an undulating one, made up of a posterior concave mandibular or glenoid fossa; the posterior slope, crest and preglenoid plane of the articular tubercle (articular eminence), and the entoglenoid process (the pointed tip of the squamous temporal above the spine of the sphenoid) medial to it. The mandibular condyle surmounts a narrow neck. It is twice as wide as from front to back. It is slightly tilted forwards. Its long axis is at right angles to the body of the mandible and if projected medially it passes slightly backwards to the anterior border of the foramen magnum.

ii. the **covering of the bone surfaces** in the adult comprises a deep layer of fibrocartilage overlain by avascular dense fibrous tissue. In the immature mandible the deepest layer is of proliferative hyaline cartilage (see question 489).

iii. the **capsule** is attached firmly to the neck of the mandible, but loosely around the cranial articular surface.

iv. the joint is divided by a moulded, complete, oval, fibrous **disc**, 1 mm thick in the middle, 2 mm thick in front and 3 mm thick behind. It is attached all round to the capsule and is fixed to the pointed ends of the condyle. The thick posterior part of the disc is made up of an upper and lower layer separated by spongy vascular elastic connective tissue containing many nerve endings. The lateral pterygoid tendon partly inserts into the disc anteriorly.

v. the **temporomandibular ligament** arises from the lateral surface of the articular tubercle. Its fibres pass downwards and backwards to the back of the neck of the mandible. Deep fibres of the ligament blend with the capsule. It prevents excessive retraction of the condyle against the tympanic plate.

vi. two adjacent ligaments are often described as accessory to the temporomandibular joint. The **sphenomandibular ligament** extends from the spine of the sphenoid to the lingula of the ramus of the mandible. However, tension in the ligament does not vary significantly during opening and closing of the joint. The **stylomandibular ligament** extends from the lateral side of the styloid process to the lower posterior border of the mandible and its angle. It is the anterior edge of the deep cervical fascia medial to the parotid gland, which separates the parotid from the submandibular gland. It is lax in both the open and closed position of the mandible and is tensed only by maximal protrusion of the jaw. Neither ligament is now considered to play any part in controlling movements of the mandible.

480. Where are the muscles of mastication inserted into the mandible ?

Lateral pterygoid is inserted into the pterygoid fovea of the front of the neck, and to the articular capsule and disc.

Medial pterygoid is inserted into the medial surface of the lower part of the ramus, outlined by the mandibular foramen, posterior border, angle, base and nearly as far forward as the groove for the nerve to mylohyoid. The area may be roughened by tendinous fasciculi.

Temporalis is inserted by two tendinous laminae. The lateral lamina inserts into the medial surface, posterior border, apex and anterior border of the coronoid process, and the uppermost part of the anterior border of the ramus as it becomes the oblique line. The medial lamina inserts into the temporal crest, the sharp margin medial to the retromolar fossa.

Masseter is inserted into almost the whole of the lateral surface of the ramus, except its upper posterior part where it is covered by parotid gland. The surface shows tendinous ridges, and the angle may be slightly everted, presumably due to the pull of the lower fibres.

481. What movements occur at the temporomandibular joint ?

The movements that occur at the temporomandibular joint are —

Depression, or opening, in which lateral pterygoid pulls the condyle and articular disc forwards. Separated by the disc, the condyle rolls and slides downwards and forwards on the articular tubercle, onto the preglenoid plane. In the greater part of this movement the mandible rotates around a transverse axis near the mandibular foramina. It may be that in the first few degrees of opening there is rotation about the heads of the mandible before anterior sliding occurs.

Elevation, closing or occlusion, in which the reverse occurs.

Protrusion, in which the condyle moves forwards without depression of the mandible.

Retraction, which is the reverse of protrusion.

Lateral, or chewing, in which there is alternate protrusion and retraction of the two joints.

Most stress through the joint is taken on the articular tubercle. The floor of the mandibular fossa is quite thin. It is translucent in the dry skull when held up to the light. Most of the pressure of occlusion is taken by the molar teeth.

482. What muscles move the mandible ?

The muscles that move the mandible, on each side, are —

In depression	—	Lateral pterygoid, digastric, and geniohyoid and mylohyoid from a stabilised hyoid bone. Gravity plays a part.
In elevation	—	Temporalis, masseter, and medial pterygoid.
In protrusion	—	Lateral and medial pterygoid.
In retraction	—	Posterior fibres of temporalis, digastric, and genio-hyoid and anterior mylohyoid from a stabilised hyoid.

483. What muscles of the mouth are attached to the mandible ?

Genioglossus is attached to the superior genial tubercle.

Geniohyoid is attached to the inferior genial tubercle.

Mylohyoid is attached to the mylohyoid line.

Buccinator arises from the outer surface of the alveolar process from the retromolar triangle to level with the first molar tooth (see also question 469).

Depressor anguli oris arises from the oblique line below the canine and first and second premolar teeth.

Depressor labii inferioris arises from the oblique line deep to the anterior part of the depressor anguli oris forward and medial to the mental foramen.

Mentalis arises from below the lateral incisor tooth.

484. Identify the mandibular foramen. What passes through it ?

The mandibular foramen is in the middle of the inner surface of the ramus. It marks the beginning of the mandibular canal. Its anterior border is enhanced by a thin, backwardly projecting flange of bone called the lingula. A narrow mylohyoid groove runs downwards and forwards from the lower posterior part of the foramen. A wider shallow groove also passes upwards and backwards from the foramen towards the neck. The sphenomandibular ligament expands fan-like at its insertion into the lingula, and below the foramen round to the lower end of the groove above it. The inferior alveolar nerve and vessels approach the foramen from above and medially, between the ramus and the sphenomandibular ligament. Before passing through the foramen they give off mylohyoid branches which penetrate the ligament to run in the mylohyoid groove. During opening and closing of the jaw the foramen moves up and down a few millimetres, but does not move appreciably anteriorly or posteriorly. The foramen, and inferior alveolar nerve, can be approached anteriorly from the opposite side of the open mouth, by a needle passed backwards on a level with the occlusive surfaces of the lower teeth, medial to the bulge of the deep tendon of temporalis, between it and the anterior border of the medial pterygoid. If the mouth cannot be opened the nerve can be approached from below along the middle of the deep surface of the ramus, from in front of the angle of the jaw.

485. Identify the mental foramen. What passes through it ?

The mental foramen lies below the second premolar tooth or between the second premolar and first molar, and midway between the upper and lower borders, or a little lower. In the newborn it faces forwards, but in infancy, as the chin develops it comes to face upwards, and then upwards and backwards. The mental nerve and vessels pass through it. A few drops of local anaesthetic injected by a needle passed into the canal will anaesthetise the premolar, canine and incisor teeth.

486. What nerves are in contact with the mandible ?

The **lingual nerve** is in contact with the anterior, inner surface of the ramus, and the lingual surface of the alveolar process from the first to the third molar tooth, deep to the mucous membrane.

The **inferior alveolar nerve** passes into the mandibular foramen.

The **mental nerve** emerges from the mental foramen.

The **auriculotemporal nerve** winds posteriorly round the neck of the bone.

The **nerve to masseter** passes laterally anterior to the temporomandibular joint above the lateral pterygoid, and behind the coronoid process and temporalis.

The **marginal mandibular branch of the facial nerve** may cross the lower border of the bone at the lower anterior attachment of masseter (where also the facial artery passes onto the face, and where there may be a lymph node).

487. What is the nerve supply of the teeth of the mandible ?

The nerve supply to the pulp cavities and deep dentine of the teeth of the mandible on each side is very probably entirely by the **inferior alveolar nerve**. It enters the mandibular foramen to pass in the mandibular canal as far as the mental foramen. Here it gives off the mental nerve, which passes through the mental foramen and then continues on in a small canal as the incisive nerve. The mandibular canal varies in position. It may be quite low, towards the base of the mandible, giving off relatively long branches to the molar and premolar teeth, or it may be closely related to the roots giving off short branches. Intermediate positions of the canal are common. Terminal branches of the nerve exchange fibres in a loose plexus below the roots of the teeth. From this plexus, nerve fibres pass to the periodontal membranes and gum, and into the root canaliculi.

In the lower jaw **additional** nerve supply reaches the periodontal membranes via numerous small branches passing through neurovascular canals from the surface of the bone on both sides of the alveolus, but there seems to be no conclusive evidence that these fibres actually pass into the pulp cavities.

On the buccal side of the alveolus these additional branches arise from —
An incisive plexus, formed from the mental nerve. They pass to the region of the first premolar, canine, and incisor teeth.
Branches of the buccal nerve. They pass to the region of the second premolar and molar teeth.
Branches from nerves to the temporalis, through the retromolar fossa, and to the masseter, through its insertion. They pass to the molar teeth.

On the lingual side of the alveolus additional branches include those derived from a smaller incisive plexus, formed similarly to the buccal plexus by the mental nerve, and those derived from the lingual nerve. In about 10% of subjects a terminal branch of the nerve to mylohyoid penetrates the mandible in the mental region and is sensory to the incisor teeth. Branches from the incisive plexuses may reach across the midline to the opposite incisor teeth.

Anaesthesia of the teeth for the drilling of cavities usually requires block of the inferior alveolar nerve only. Though it has been said that in 5% of patients this is not totally effective, it is not certain that this is due to the presence of additional nerve supply. For more severe manoeuvres that involve periodontal nerve endings, however, block of surface alveolar nerves as well as of the inferior alveolar nerves is required. Usually the outer plate of the alveolus in the region of the incisor and canine teeth is thin enough to allow diffusion across it of local anaesthetic, injected into the overlying gum, to be effective. The outer plate over the premolar and molar teeth is too thick to allow surface anaesthesia to be effective by itself, so that in these circumstances block of the lingual and or buccal nerves is also carried out.

488. What salivary glands are in contact with the bone?

The **parotid gland** rests against the posterior margin of the ramus, between the insertions of the masseter and medial pterygoid muscles. It encroaches onto the back and lateral aspects of the neck of the bone.

The **submandibular gland** rests against the submandibular fossa of the bone, outlined by the mylohyoid line and the inferior border or base of the bone, as far back as the insertion of medial pterygoid into the ramus.

The **sublingual gland** rests against the sublingual fossa of the bone, above the anterior part of the mylohyoid line.

489. How does the mandible develop ?

The mandible develops in membrane lateral to the inferior alveolar nerve and its incisive branch, and to the lower portion of Meckel's cartilage. The centre of ossification of each half of the bone is the second to appear in the body, after that for the clavicle. Meckel's cartilage and its perichondrium become surrounded by the developing bone and disappear apart from the lingula, sphenomandibular ligament, anterior malleolar ligament, malleus and incus. The developing teeth become surrounded by bone. At birth the mandible is in two separate halves united at the symphysis. This fuses during the first year. The surface cartilage of the condyle acts as an epiphyseal plate. The remaining bone remodels with growth.

490. What changes occur in the mandible in old age ?

As teeth are lost, the alveolus is absorbed, leaving an alveolar ridge. This is derived from the thicker buccal margin of the alveolus over the molar region and the lingual margin of the alveolus over the canine and incisor region. The arch of what remains of the alveolus is thus widened posteriorly. Also, as atrophy of the mandible progresses, the upper border of the body can sink to the level of the genial tubercles, and the mylohyoid lines, to the attachments of the genioglossi, mylohyoids and buccinators. This makes the fitting of dentures difficult. In extreme atrophy the upper wall of the mandibular canal can become very thin or even deficient, exposing the inferior alveolar nerve to pressure. The ramus becomes a little more oblique to the body and the neck slopes backwards.

491. What does the hyoid bone do ?

The hyoid bone —
  i. acts as a mobile but firm base for muscles of the tongue, floor of the mouth and pharynx. Its position can be adjusted by the pull of muscles in opposite directions, upwards and forwards to the jaw by the geniohyoids, backwards and upwards to the cranium by the stylohyoids and downwards to the sternum by the sternohyoids. From a stabilised hyoid bone the tongue muscles (hyoglossus and chondroglossus), the muscle of the floor of the mouth (mylohyoid) and the pharyngeal muscle (middle constrictor) can function more effectively. It is therefore important for chewing, swallowing and speech.
  ii. suspends the larynx, and via its muscles gives it more mobility.
  iii. protects the front of the lower pharynx and the opening of the airway.

SOURCE BOOKS, most of which contain extensive references to the specialised Journals.

Backhouse K.M. and Hutchings R.T. A Colour Atlas of Surface Anatomy. Wolfe Medical Publications Ltd. 1986.

Basmajian J.V. Muscles Alive (fourth edition). Williams and Wilkins, Baltimore. 1978.

Bogduk N. and Twomey L.T. Clinical Anatomy of the Lumbar Spine. Churchill Livingstone. 1987.

Bowers W.H. The Interphalangeal Joints. Churchill Livingstone. 1987.

Breathnach A.S. Frazer's Anatomy of the Human Skeleton. Churchill Ltd.(fifth edition — unfortunately out of print). 1958.

Cormack C.C. and Lamberty G.H. The Arterial Anatomy of Skin Flaps. Churchill Livingstone. 1986.

Cormack D.H. Ham's Histology (ninth edition). J.B. Lippincott Co. 1987.

Crenshaw A.H. Campbell's Operative Orthopaedics (seventh edition). C.V.Mosby Co. 1987.

Dandy D.J. Arthroscopy of the Knee. A diagnostic Colour Atlas. Gower Medical Publishing. 1984.

Doxanas M.T. and Anderson R.L. Clinical Orbital Anatomy. Williams and Wilkins. 1984.

DuBrul E.L. Sicher's and DuBrul's Oral Anatomy (eighth edition). Isbiyaku EuroAmerice, Inc. 1988.

Fawcett D.W. Bloom and Fawcett. A Textbook of Histology. W.B. Saunders Co. 1986.

Ghadially F.N. Fine Structure of Synovial Joints. Butterworths. 1983.

Hollinshead W.H. Anatomy for Surgeons. Vols. 1 and 3 (third edition) 1982. Vol. 2 (second edition) 1972. Harper and Row.

Inman V.T. The Joints of the Ankle. Williams and Wilkins. 1976.

Kapandji I.A. The Physiology of the Joints, Vols. 1–3. Churchill Livingstone. 1970.

Keogh B. and Ebbs S. Normal Surface Anatomy, with Practical Applications. William Heinemann Medical Books Ltd. 1984.

Kelikian H. and Kelikian A.S. Disorders of the Ankle. W.B. Saunders Co. 1985.

Lierse W. Applied Anatomy of the Pelvis. Springer-Verlag. 1984.

Lundborg G. Nerve Injury and Repair. Churchill Livingstone. 1988.

Mackinnon S.E. and Dellon A. Lee. Surgery of the Peripheral Nerve. Thieme. 1988.

McMinn R.M.H. Last's Anatomy, Regional and Applied. Churchill Livingstone. 1990.

McMinn R.M.H., Hutchings R.T. and Logan B.M. A Colour Atlas of Applied Anatomy. Wolfe Medical Publications Ltd. 1984.

Milford Lee W. Retaining Ligaments of the Digits of the Hand. W.B.Saunders Co. 1968.

Muller W. The Knee. Form, Function, and Ligament Reconstruction. Springer-Verlag. 1983

Rickenbacher J., Landolt A.M. and Theiler K. Applied Anatomy of the Back. Springer-Verlag. 1982.

Rockwood C.A. Jr. and Green D.P. Fractures in Adults. Vols. 1 and 2 (second edition). J.B. Lippincott Co. 1984.

Rockwood C.A. Jr., Wilkins K.E. and King R.E. Fractures in Children. Vol. 3. J.B. Lippincott Co. 1984.

Salasche S.J., Bernstein G. and Senkarik M. Surgical Anatomy of the Skin. Appleton and Lange. 1988.

Sarrafian S.K. Anatomy of the Foot and Ankle. J.B. Lippincott Co. 1983.

Skandalakis J.E., Gray S.W., Mansberger, A.R. Jr., Colborn G.L. and Skandalakis L.J. Hernia, Surgical Anatomy and Technique. McGraw-Hill Information Services Co. 1989.

Spinner M. Injuries to the Major Branches of the Peripheral Nerves of the Forearm (second edition). W.B. Saunders Co. 1978.

Spinner M. Kaplan's Functional and Surgical Anatomy of the Hand (third edition). J.B. Lippincott Co. 1984.

Sunderland Sir S.S. Nerves and Nerve Injuries. Churchill Livingstone. 1978.

Taleisnik J. The Wrist. Churchill Livingstone. 1985.

Tubiana R. The Hand. Vol. 1. W.B. Saunders Co. 1981.

Warwick R. Eugene Wolff's Anatomy of the Eye and Orbit. H.K. Lewis and Co. Ltd. 1976.

Williams P.L., Warwick R., Dyson M. and Bannister L.H. Gray's Anatomy (thirty-seventh edition). Churchill Livingstone. 1989.

Zancolli E. Structural and Dynamic Bases of Hand Surgery (second edition). J.B. Lippincott Co. 1979.